Simulating Ecological and Evolutionary Systems in C

Many recent advances in theoretical ecology and evolution have been made by bringing together biological, mathematical, and computational approaches, yet there are very few books available that contain this particular mix of information. This book is one of the first to consider all three approaches in one volume, using the widely available computer programming language C and biologically motivated individual-based simulations involving processes such as competition, foraging, predation, mating systems, and life-history optimization. All of the important features of C are covered, providing an excellent resource for those seeking to adopt a computational approach.

Simulating Ecological and Evolutionary Systems in C

Will Wilson
Department of Zoology, and
The Center for Nonlinear and Complex Systems
Duke University

CAMBRIDGE
UNIVERSITY PRESS

PUBLISHED BY THE PRESS SYNDICATE OF THE UNIVERSITY OF CAMBRIDGE
The Pitt Building, Trumpington Street, Cambridge, United Kingdom

CAMBRIDGE UNIVERSITY PRESS
The Edinburgh Building, Cambridge, CB2 2RU, UK http://www.cup.cam.ac.uk
40 West 20th Street, New York, NY 10011–4211, USA http://www.cup.org
10 Stamford Road, Oakleigh, Melbourne 3166, Australia
Ruiz de Alarcón 13, 28014 Madrid, Spain

First published 2000

Printed in the United Kingdom at the University Press, Cambridge

Typeface Computer Modern 11/14pt *System* LaTeX [UPH]

A catalogue record of this book is available from the British Library

Library of Congress Cataloguing in Publication data
Wilson, Will, 1960–
Simulating Ecological and Evolutionary Systems in C / Will Wilson.
p. cm.
Includes bibliographical references and index.
ISBN 0 521 77228 1 (hb)
1. Ecology–Computer simulation. 2. Evolution (Biology)–Computer simulation.
3. C (Computer program language). I. Title.
QH541.15.S5 W54 2000
577′.01′13–dc21 99-087678 CIP

ISBN 0 521 77228 1 hardback
ISBN 0 521 77658 9 paperback

To Frances, Hannah, and Ella

Contents

Preface

This book arises from a course I've taught at Duke University in the Department of Zoology. The goal of the course, like the goal of this book, is to combine mathematical and simulation approaches within a common framework to pursue interesting questions in population biology. The book's perspective reflects my interest in linking population-level phenomena and individual-level interactions. I address upper-level undergraduates, graduate students, and researchers familiar with theoretical concepts in ecology and having a strong mathematical foundation.

The main challenge in the book (which is heightened in a one-semester course) is that three concurrent goals must be addressed simultaneously. One goal is teaching the C language to students who may have never programmed in any language. Another goal is teaching new mathematical concepts to students, in particular the mathematics associated with analyzing spatial problems, as well as standard approaches such as the stability analysis of a nonspatial model. A third goal is using simulation and mathematical approaches together to understand ecological dynamics by comparing and contrasting deterministic and stochastic modeling frameworks.

These three concurrent goals are reflected in the book, which serves much like a workbook for students learning C and applying it to ecological and evolutionary problems. Most of the theoretical tools I've needed in my work – programming concepts and mathematics – are mentioned somewhere in the book, along with useful references having more complete treatments. All of the code shown in the text, including extensions, is available for downloading from my website (presently http://www.zoology.duke.edu/wilson/).

In teaching this course I have had some students with no programming experience, others with no background in ecological theory, and some with neither, but I have excluded students without mathematics experience. The resulting emphasis of course time is always somewhat frustrating: About

20% is spent discussing ecological problems and their traditional mathematical approaches, 60% teaching C and subsequent simulations, and 20% examining analytic models that provide better descriptions of the simulation results. The heavy emphasis on programming can be frustrating, at least to me, because I believe that the best simulation is one that makes itself obsolete by replacement with an analytic formulation. Although such a replacement is not always possible, there is much work being performed using simulation models that have no connection whatsoever to the strong foundation of theoretical ecology. One can not leave students with the impression that the act of writing a program that simulates an ecological system advances theoretical ecology. An aspect that tempers my frustration is that I don't believe the best program is the one that is unwritten. Historically, much theoretical ecology has been performed in mathematics departments, and in some cases the thought of performing a computer simulation seems to be heretical. Much work in the past decade or two shows that this dogmatic attitude is changing. My point here is that mathematical models untested by simulations are roughly equivalent to simulations ungrounded by mathematical theory. It is time the two approaches are unified under one theoretical motivation.

I emphasize developing a personal programming style – of using white space and writing comments – as you gain more programming experience. Such details don't matter to the compiler and the central processing unit of a computer, but having cleanly written code helps while reexamining long-forgotten code. I'm a scientific programmer, hence, the programs shown here are not written to the standards of a computer scientist. As such, my code might not be pretty, and may sometimes be an inefficient way to get the job done, but it emphasizes the structure of the biological problem involved.

In many ways, computer languages are interchangeable when trying to think about a scientific problem. I started learning BASIC in ninth grade, learned FORTRAN while an undergraduate, took my only college programming class in PASCAL, and learned C while a postdoctoral researcher. At present I only use C. One never wants to invest valuable time learning a dying language: C is very much alive and well. It serves as the foundation of the Unix operating system (and Linux too), which will not perish for at least decades. Thus, the initial time investment should pay off in flexibility, stability, and utility. I'm also a firm believer of Mangel and Clark's (1988, page 7) credo: "Learn new things when you need them." Hence, I use many programming and system-level features as black boxes, especially when it comes to hardware and operating systems. I am hesitant to learn new tricks in programming, such as C++, object-oriented programming, development packages, and even debuggers. I wouldn't go so far as to say those things aren't worth learning, but I want to maximize time spent doing fun stuff like science, not keeping up with the continuous stream of new and improved software development tools.

I thank many people. My first computer programming (and physics) teacher was Gene Scribner, whose attention and patience began my academic pursuits. I am particularly indebted to Peter Crooker, Lawrence Harder, Bill Laidlaw, Ed McCauley, Roger Nisbet, Kris Vasudevan, and Chester Vause for guiding me through a number of scientific disciplines. Many others have influenced me: Spencer Barrett, Niko Bluethgen, Kim Cuddington, Andre de Roos, Randy Downer, Steve Ellner, Mike Gilchrist, Lloyd Goldwasser, Susan Harrison, Peter Kareiva, Simon Levin, Craig Osenberg, Mark Schildauer, Doug Taylor, Kris Vasudevan, Steve Vogel, Kim Wagstaff, and my colleagues at Duke University. Bill Morris, Frances Presma, Anne Rix, Kris Vasudevan, and Bill Laidlaw gave very useful comments throughout the manuscript. Many others gave detailed comments on specific chapters, including David McCauley, Hugh Possingham, Shane Richards, Colette St. Mary, and Peter Turchin.

Durham, North Carolina W.G. Wilson
January 2000

1 Introduction

Programming components covered in this chapter:
- operating systems
- essential programming concepts
- compilers
- personal progamming

1.1 Theory, Numerical Methods, and Simulations

Organisms of one or more species within a common environment constitute an ecological system. Their mutual interactions lead to births and deaths, consumption and growth; these changes represent the ecological system's dynamics. Empiricists study real systems, either in the field or in the laboratory, whereas theorists study idealized systems using a variety of methods.

When either type of ecologist summarizes the most important theoretical advances in ecology, it is likely that the list contains concepts arising from simple analytic models such as the Lotka–Volterra (LV) predator–prey model and competition models (e.g., Hastings 1997), the SIR (Susceptible–Infected–Recovered) model of infectious diseases (e.g., Murray 1989), and Levins's (1969) metapopulation model. Although simplistic caricatures of real ecological processes, it is their simplicity that makes their conceptual implications all the more powerful. The goal of all *theory* is conceptual advance within a particular area of research.

One criticism of simplistic analytical models is their inability to capture the subtleties of ecological reality. Complications lead to narrow limits and broad qualifications of all general conclusions when a simplistic model is applied to specific situations. For example, no one can reasonably expect the Lotka–Volterra predator–prey model to describe the observed dynamics of the North American lynx–hare oscillation over the last few centuries

(Gilpin 1973). Understanding a specific system often requires specific assumptions about life histories, environments, and the interactions with additional species – assumptions that greatly qualify the applicability of general models and may render their predictions impotent. Taking account of these specific assumptions often makes an analytic formulation intractable, meaning that no carbon-based life form, or even a silicon-based one, can extract a meaningful, analytic solution.

How do you solve an analytically unsolvable model? Over the last few decades the answer has been to throw a computer at the model, sometimes literally, and find numerical solutions to messy equations. Many people have devoted much time and energy producing tremendously useful programs designed to solve mathematical problems: Maple and Mathematica, for example, as well as specific programs within theoretical ecology itself, such as the Solver program for time-delayed ecological problems (Gurney and Nisbet 1998). All these approaches fall within my definition of *numerical methods*, where the concepts are first laid out within an analytic framework, followed quickly by a search for a method to reach a solution. The computer is used as a very refined calculator.

This book emphasizes using computers as a different kind of theoretical tool – programming a computer to *simulate* ecological systems containing many individuals that interact stochastically. Several recent reviews cover many examples of individual-based simulation work (Hogeweg 1988, Huston, DeAngelis, and Post 1988, Uchmański and Grimm 1996, Grimm 1999). There are two main advantages that simulations of ecological and evolutionary systems have over analytic approaches. First, simulation models can incorporate an arbitrary amount of complicated, biologically realistic processes, for example, age- and size-dependent processes and experience-dependent individual-level decisions. This ability mitigates concerns that models are too simplistic and unrealistic. Second, these complicated processes can incorporate the stochasticity, or the randomness, inherent to biological interactions; for example, in which direction does an individual take its next step? These simulations can become so detailed that many users of these models abandon analytic formulations in favor of biological realism (e.g., Schmitz and Booth 1997).

However, I do not argue for replacing mathematical models with computer simulations – analytic models are the best encapsulation of ecological and evolutionary mechanisms. I favor an approach based on the assertion that comparing multiple models of an ecological system yield theoretical insights unattainable from a single model (e.g., McCauley, Wilson, and de Roos 1993), an approach which may reflect a general trend in the field (Grimm

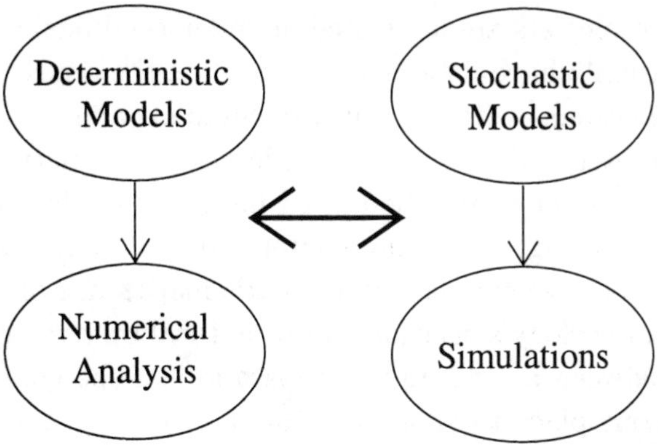

Figure 1.1. Examining deterministic models often requires numerical solutions, whereas stochastic models require computer simulations. Interplay between deterministic and stochastic models can help develop conceptual insight.

1999). Analytic models profess to describe empirical ecological systems. It makes sense to test the conclusions of these models against the output of a computer program that simulates, in an idealized and controlled manner, one of these empirical systems. Why? First, maybe the theorist forgot a fundamental process or assumed that one thing was not important whereas another was. Judicious use of a simulation can check the importance of various processes and the validity of assumptions. Often, simulation results help refine model formulations (see figure 1.1) so much so that, in the end, the essential processes look nothing like the initially proposed analytic ones. It is not an issue of one theoretical tool being "better" than another, rather it is an issue of using two tools together, much like the ideal link between theoretical and empirical pursuits (Caswell 1988).

The idea of simulating ecological systems can be traced back to Lotka (1924),[1] when he suggested putting theoretical ecologists around a game board, playing such roles as predator and prey individuals, to simulate the interactions of an ecological system. His goal of ecological game-playing was to provide insight into population-level dynamics. It is around Lotka's motivation that the computer programs in this book are designed. The computer is an ideal machine to keep track of all the many individuals being

[1] The practice of scientific simulations can be traced back to Metropolis *et al.* (1953) in physics and Bartlett (1955) for general stochastic processes.

played in the game, update their interactions, measure their population-level averages, and visualize their collective dynamics.

However, ecologists are interested in understanding natural ecological systems, and usually have little desire to replace the study of beautifully complex natural systems with the study of complex, artificial systems generated on the computer. The desire to understand the natural world demands that these computer results be placed into the broader, analytic framework of theoretical ecology. It is my contention that doing so produces insight unavailable to either simulations or mathematics in isolation.

Theoretical ecology's relationship to natural systems is much like the relationship between a map and a landscape.[2] If the goal in using a map is to get from one place to another within a city, then important details such as side-streets and landmarks might be helpful, but picky details like the position of every building and tree are annoying distractions. Alternatively, if the goal is to move from one place to another across a continent, then the locations of side-streets and minor roads become annoying distractions. A map with too many details subverts the original need for the map – if you feel that all details must be included, then you just reproduce the landscape without any synthesis of the landscape's information.

A simulation model is like a detail-rich map – it represents an idealized ecological system encapsulating the important interactions between organisms, but not so many that the model becomes useless. The ecologist's job is to pinpoint the mechanisms, or the key processes, that determine the resultant patterns and should therefore be included in a map with less detail. This job requires understanding and synthesizing mountains of simulation data and to find the appropriate analytic model that makes a detailed simulation details expendable. Linking the individual interactions to population-level models is one important goal of ecological theory.

I hope to demonstrate the use of simulations in pursuing an understanding of ecological systems. The entire process of programming is geared to constructing a logical set of rules for a particular task. Writing the final code simulating an ecological system demands clear and concise thoughts about the important features of the system. Of course this procedure is iterative, and the first attempt at a particular program is a crude characterization of the final product. The entire creation process is much like performance art – the numerical results are of lesser importance than the conceptual development that takes place in the thoughts of the creator.

[2] An analogy I first read in work of Kim Cuddington's.

I firmly believe that the more models you have of a particular ecological process, the better. In this book, the ecological concepts are usually first presented in terms of preliminary analytic encapsulations, then the implied individual-scale rules are translated into C code. Connections between models are made through comparisons of simulation results with those of the analytic model(s). In a few of the chapters I have only just begun to explore the connections, even though potential analytic models of more detailed simulations might be sitting there ready to be solved! But there is much work left to be done with many problems in ecological theory – that situation makes it a fascinating area of research.

1.2 An Example System: Predator–Prey Interactions

One of the most famous and simplest models in theoretical ecology is the Lotka–Volterra model of predator–prey interactions. Useful descriptions can be found in most introductory texts discussing ecological theory. The interactions between a species, the prey, that is a resource for a second species, the predator, are described. Although it is one of the oldest characterizations of predator–prey dynamics, it contains many biologically implausible features that prevent its application to specific, real ecological systems. Yet the Lotka–Volterra model is pedagogically useful in thinking about the interaction of species, and we will perpetuate its use here as an example of connecting deterministic and stochastic models.

Assumptions. Imagine a microbial system of prey and predators continuously stirred on a Petri dish, or in a beaker, to prevent the generation of patchiness. There are four basic interactions. First, prey reproduce clonally at a density-independent rate α (alpha). Second, predators encounter, attack, and consume prey with a rate β (beta). Third, when a prey item is consumed, its biomass is converted into new predators with efficiency ϵ (epsilon). Finally, predators die with rate δ (delta).

1.2.1 Ordinary Differential Equation Model

Often there is an excellent correspondence between the dynamics of a collection of many interacting, discrete entities and the solution of a set of ordinary differential equations, evidenced by work in many fields including population dynamics, chemical reaction kinetics, and hydrodynamics to name just a few. The assumed interactions listed above describe the following set of

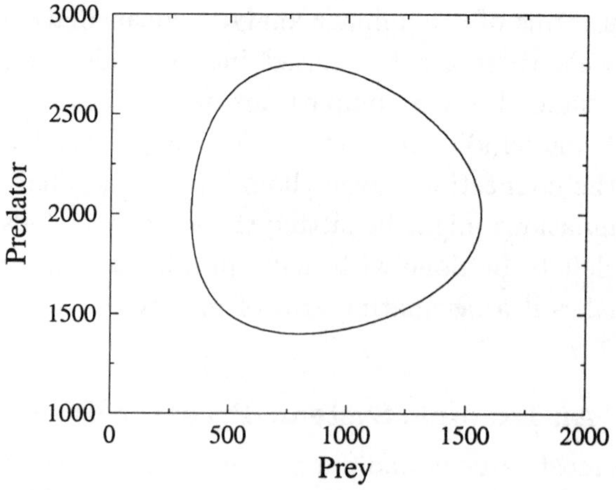

Figure 1.2. Exact solution to the Lotka–Volterra predator–prey model in phase space (see equation (1.3)). ($\alpha = 0.5$, $\beta = 1.0$, $\epsilon = 0.5$, $\delta = 0.1$, $N = 4000$, $V_0 = 800$, $P_0 = 1400$)

ordinary differential equations (ODEs)

$$\frac{dV}{dt} = \alpha V - \frac{\beta}{N} PV \tag{1.1a}$$

$$\frac{dP}{dt} = \epsilon \frac{\beta}{N} PV - \delta P, \tag{1.1b}$$

where V and P are the prey (V for victim) and predator densities (e.g., number per meter squared), respectively. The first equation describes temporal changes in the prey density, with the first term representing prey reproduction and the second term representing prey removal by predation. The second equation describes predator dynamics with the first term representing conversion of consumed prey into new predators and the second term representing predator mortality. The parameter N represents the relative scale of the number of organisms in the system such that, for example, V/N is the relative liklihood that a predator meets a prey on any given search for food.

Even for such a simple set of mathematical equations as the above Lotka–Volterra predator–prey model it is not possible to obtain an exact time-dependent solution by analytic methods. However, we can find an analytic solution relating the two species densities to one another (see Murray 1989).

If (1.1a) is divided by (1.1b), we obtain the differential equation

$$\frac{dV}{dP} = \frac{V(\alpha N - \beta P)}{P(\epsilon\beta V - \delta N)} \tag{1.2}$$

which can be solved to yield

$$\alpha N \ln\frac{P}{P_0} + \delta N \ln\frac{V}{V_0} = \epsilon\beta(V - V_0) + \beta(P - P_0), \tag{1.3}$$

where V_0 and P_0 are the initial prey and predator densities (see page 237). Thus, given the prey density, we can calculate the predator density,[3] but we cannot calculate the times at which these densities occur. Figure 1.2 shows the prey–predator densities as a phase plot (dynamical variables plotted against one another) determined numerically from (1.3). The curve is a closed cycle, indicating the cyclic nature of the Lotka–Volterra predator–prey model. This model's cycle is called neutrally stable because the numerical values of the cycle depend on the initial conditions, whereas a stable cycle (called a limit cycle) would be independent of initial conditions.

1.2.2 Simulation Models

Another way of exploring the assumed predator–prey interactions is through a brute-force simulation of prey and predator individuals. Suddenly, with this route, there become many ways to translate the explicitly stated set of predator–prey interactions into simulation rules because representing the individuals and their interactions within a computer program brings up many questions in need of resolution. Are the individuals point-particles (and can be packed with infinite density) or do they take up space? Another way of asking this question is, "does each individual interact with all other individuals within an infinitesimal time Δt?" These two questions are related if interaction rates between individuals are somehow dependent on their separation. Both options can lead to a simulation that matches the predictions of the ODE model, but the differences in their detailed assumptions provide distinct foundations for model extensions.

One translation of the assumptions into simulation rules, depicted in figure 1.3, assumes a discrete-time updating of prey and predator populations scattered over a lattice of cells. These rules list cell states before an interaction on the left-hand side, the probability that the interaction occurs during a time Δt above the arrow, and the cell states resulting from the interaction

[3] Note that there are two values for each prey density, and similarly for the predator density.

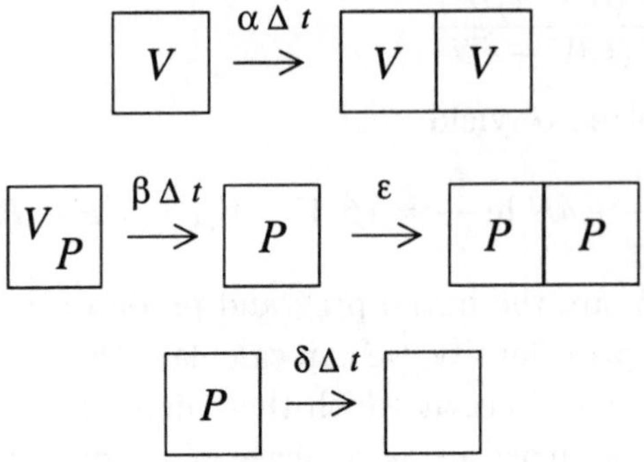

Figure 1.3. Simulation rules corresponding to the Lotka–Volterra predator–prey interactions encapsulated by equation (1.1). The boxes represent lattice cells and Ps and Vs represent occupation by a predator and/or prey. Arrows represent interactions, occurring with the probabilities listed above them, that alter cell occupation states. The top set represents prey reproduction, the middle set represents predation and subsequent predator reproduction, and the bottom set represents predator death.

on the right-hand side. Among the many simulation assumptions is that individuals take up the space of about one cell and, therefore, cells can contain only one prey and/or one predator at a time. During a very short time interval Δt an individual interacts only with nearby individuals, and to maintain spatial homogeneity the locations of individuals must be randomized at a rapid rate.

Consider, as an example, the interactions associated with predation. If a prey and predator are found in the same cell then the prey is consumed with probability $\beta\Delta t$,[4] otherwise nothing happens. If predation takes place, then predator reproduction occurs with probability ϵ, but because cells are limited to one of each species, the offspring is placed into a neighboring cell.

1.2.3 Connections between Models

Relating simulation and analytic formulations arising from a common set of ecological assumptions is what this book is about. Analytic formulations are always an ideal theoretical goal, but testing the many assumptions of an

[4] If the time interval Δt becomes long, then the interaction probability is more accurately described by $1 - \exp(-\beta\Delta t)$. Another way of formulating simulations, called *discrete-event simulations*, makes extensive use of exponentially distributed event times.

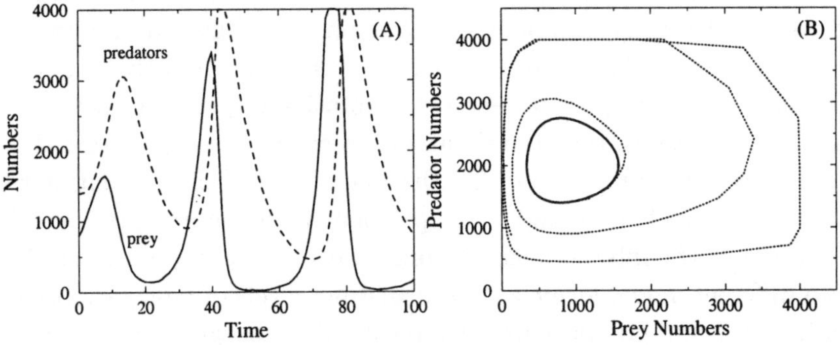

Figure 1.4. Results from a Lotka–Volterra predator–prey simulation implementing the rules of figure 1.3. (A) The amplitude of the population oscillations increase rapidly with time. (B) Plotting the simulation results (thin line) in phase space and comparing with exact analytic expectations (thick line) reveals important differences. An understanding of the sources for such differences is the benefit of comparing simulation and analytic models.

analytic framework can be dealt with directly and easily using an individual-based simulation. The simulation is not intended to be a replacement for clearer and more generalizable mathematical descriptions, but rather an essential part of connecting real ecological systems to a mathematical formulation.

The simplest conceptual linkage between the stochastic simulation model and the deterministic ODE model, equation (1.1), is that the population densities V and P represent the two species' cell occupation probabilities. Likewise, the parameters are identified as the cell state transition rates.[5]

Figure 1.4A shows the temporal dynamics of population counts from a simulation implementing the above predator–prey rules (discussed in detail in chapter 9) with a simulation time step of $\Delta t = 1$. Given initial prey and predator numbers, oscillations appear in the population densities of both species, and the amplitude of the oscillations grows with each period. Plotted in the same way as the analytic results in figure 1.4B we observe an outward spiral in the prey–predator plot instead of the well-contained cycle anticipated from the analytic results.

[5] Making and proving mathematical connections between stochastic and deterministic models is a nontrivial task (e.g., Durrett and Levin 1994).

First, there is a clear lack of congruence between the results of the analytical model and the simulation. Why are they different? Which, if either, is correct? The latter question is a vague one; a scientist determines "correctness" based on what was supposed to be modeled. On the one hand, if the simulation was suppose to be a representation of processes described by the analytic equations, then the simulation is clearly wrong because the analytic results were not reproduced. On the other hand, if the simulation rules were taken as correct, then there is a problem with the analytic model's incorporation of the interactions. In some ways, this situation is ideal – having two different model formulations at distinct levels of biological organization, individuals and populations, with results that disagree. Resolving the paradox may bring you a new understanding of ecological processes because you are forced to think about what is really going on in each of the models.[6]

A second issue is that if one ever hopes to accurately model experimental systems, where the processes and interactions are likely to be unclear at best, a theorist might do well to practice predicting the outcome of artificial systems where the processes are exactly specified. Connecting multiple models of artificial systems builds up a set of experiences that can be used when the system is not exactly known. Finally, adding more complicated interactions is usually trivially accomplished within a simulation model but the resultant effects can be drastic. Often the solution to an analytic model precisely including these additional interactions will be unapproachable. In these cases use of the simulation must proceed with caution, attempting to check the simulation results against any and all available approximate analytic benchmarks.

1.3 What You Need to Learn

Below is a list of skills needed to complete the computer projects presented in later chapters. Don't be too overwhelmed: I only know enough of my favorite operating system's commands and features to get by, the text editor I use is very simple (thus not too much to remember), and I have only a rudimentary understanding of the many options for the C compiler. My excuse is that I am not interested in learning and remembering all the ins and outs of computers or keeping up with rapidly changing computer technology.[7]

[6] In the example shown here the explanation for the nonspatial differences is that one model implements continuous time and the other implements discrete time.

[7] In the past decade while I primarily used Unix, more popular operating systems have gone through three or four major upgrades, requiring much time, patience, and money. Things change in Unix (and C), too, but backward compatibility is much less of a problem.

Instead, I want to spend my time thinking about and working on scientific problems. That desire is the driving force behind this book.

The list of things you need to learn include:

(i) **How to use your computer's operating system.** You already use an operating system if you presently use a Windows, Macintosh, Unix, or any other kind of computer. In this book I assume the reader has access to a Unix system, although, since very little of the C programming material in the book is dependent specifically on Unix, a machine running MS-DOS works just as well. In section 1.4 I discuss operating systems in more detail.

(ii) **How to edit program files.** Program files are text files readable by both humans and C compilers, unlike the files produced by the compiler that run on the computer. Word processing programs, like Microsoft Word, are not ideal for the job of editing program files because they use various characters, invisible to the human using the program, that cause compilers to choke. These hidden characters specify fonts, styles, and other formatting choices – details irrelevant to computer programming. Instead, a program's text file uses only the alphanumeric and other visible characters.[8] On Unix systems, my preferred editor is Sun Microsystem's `textedit` text editor. This editor is nice and simple, with very few bells and whistles. Probably the most common and popular editor is `emacs` (and its more current incarnation, `xemacs`), available on all Unix systems. Another common editor is `vi`, which I discuss in chapter 2. I will not extensively cover editing programs in this book because there are too many and they are too specific for the operating system. Likewise, in many cases the C compiler is often packaged with a unique editing program tightly integrated into the programming environment.

(iii) **How to program.** Programming a computer to carry out a desired calculation requires careful consideration of the program's objective. The computer knows nothing, has no creativity, and does nothing other than what it is programmed to do – but it does what it is programmed to do very quickly. In section 1.5 I overview the three features that are fundamental and common to almost every computer programming language. The remainder of the book is then devoted to learning to program in an ecological context.

(iv) **How to compile your programs.** A compiler is a computer program (written by *real* programmers) that reads a program file written

[8] Word processing programs might work for program editing if the file is saved as "text only".

by a human and turns it into a machine-language program, readable only by a computer. I use both Sun Microsystem's compiler, `cc`, and GNU's[9] `gcc` – the free C compiler distributed with the Linux operating system. Although I emphasize Unix, all code was tested using the free LCC-Win32 C compiler for Windows, and the necessary precautions will be noted. All compilers should compile well-written portable code using the ANSI (American National Standards Institute) standard. Here are a couple of warnings, though. If you use a commercial PC compiler you may end up writing nonportable code implementing snazzy routines that constitute the reason you paid for the compiler.[10] Also, C++ compilers should compile C programs, but you may end up using a few C++ statements which will hang-up C compilers. In section 1.6 I provide an overview of C compilers and the origins of the programming language.

(v) **How to run your programs.** Once your program is compiled, you will have an executable file (like an `*.exe` file in Windows) that the computer can execute, or run. Running the program is a technically simple process. What comes out when the computer is running your program completely depends on what you put in the program and tell the program to output to you.

(vi) **How to visualize your data.** The numbers that come out of your program need to be plotted or listed or imaged in some way. A spreadsheet might help you out with plotting. A good, free plotting program for Unix and Linux systems is `xmgr`. I used it to produce all of the graphs in this book. Later I will provide code for generating PostScript files, especially handy for visualizing vast amounts of data generated by the simulation of spatial systems.

1.4 Operating Systems

If a computer is off, it doesn't do anything. When a computer is turned on, it loads what is called an "operating system" into its memory. The operating system is the basic (or not so basic) program that is always running, enabling the computer to process commands given to it by the computer user. All things are controlled by the operating system. Everything else is an executable program *run* by the operating system.

---

[9] GNU stands for "GNU's Not Unix" and is associated with the Free Software Foundation, a collection of programmers that write code and then give it away. Lots of neat, useful programs, including operating systems and compilers, can be found at their web site.

[10] OK, this problem of nonportable code can happen with C compilers on Unix systems, too.

The operating system determines and defines how a human, such as yourself, interacts with the computer. Computer hardware, to the extent that I care to understand it, is pretty much the same whether it be an IBM, Macintosh, or Sun. In comparison, humans are extremely variable in their needs, motivations, and even personalities, all of which affect their use of computers. Up to the mid-seventies, the main group of scientific computer users included physical scientists and mathematicians. This homogeneity, and the technology of the day, was reflected in the operating system. People interacted with the computer through punched cards, punched tape, or, if you were lucky, a teletypewriter. The late seventies brought cheap computers and operating systems to the masses with DOS (disk operating system), which was a text-based form of the operating system previous computer users were familiar with from mainframe computers. Both Apple's and Window's graphical user interface (GUI), a mouse-based visual way of interacting with a computer, have their origins traceable to a computer developed in the early 1970s by an Exxon research group in Palo Alto (Mullish and Cooper 1987). Presently, most operating systems have components of both these systems: A graphical user interface and a text-based window used to type commands.

Operating systems now seem divided into three worlds: Unix, Macintosh, and Windows. In the beginning was Unix (actually multics). Unix grew up on very large, relatively powerful computers. The operating systems were too big for tiny personal computers (PCs), hence conceptually important parts were excised as DOS. However, computer technology changed drastically during the 1980s and the 1990s – now even cheap PCs have the computing power, memory and disk space of the BIG computers of the 1970s. For my Ph.D. thesis work in the 1980s, I used a Cray Research Supercomputer – the Cray X-MP 48 (4 CPUs each with 8 MB memory) – and at the time it was *fast*. Using these computers to their fullest required special programming skills (e.g., Smith 1991), but now, full and complete Unix operating systems (called Linux) can be installed on $1000 PCs that are likely comparable in speed to a processor on that multimillion dollar Cray computer. You can buy a Linux OS for $50 in the computer section of your bookstore or get it free over the internet. Linux is wonderful, but requires a bit of experience to get up and running.[11] Hence, most PCs come with a simplified operating system preinstalled – Macintosh or Windows. Again, I assume throughout this book that you use a Unix system – most students and academics have access through their university. Much detailed

[11] Although these installations are becoming easier with each passing month, I have not yet installed one. For my installations I am eternally grateful to Randy Downer and David Gourley.

information on Unix can be found in a bookstore's computer section or on the web.

1.5 Computer Programming

In this section I present the main features of computer programming using C's syntax. A computer language provides a clear, logical way of thinking about scientific problems. Programming languages require even more precision than human languages since compilers have no tolerance for misspelings and other syntactical error. Although such mistakes are an annoyance for humans while reading, humans can generally understand the intended meaning by correcting the errors given the context in which they occur. Compilers, on the other hand, have no ability to correct such mistakes. This constraint of logical thought is one reason I use programming to help me formulate my conceptions of ecological processes. However, programs can still "work" with many logical flaws, which is one important reason to supplement them with mathematical models.

The underlying logic of all computer languages is based on three important statements:

(i) The = statement
(ii) The `if` statement
(iii) The `while` loop

whereas most other statements are elaborations of these. Additionally, C has some particularly useful and novel concepts that will be introduced in later chapters.

1.5.1 The = Statement

The = ("equal") statement is most appropriately called an assignment operator. Unlike in mathematics, here the equal sign has a temporal component making it very different from defining an equality: The right-hand side represents "past" values whereas the left-hand side represents the "present" value. Consider,

```
x =x + 1;
```

This statement is a mathematical absurdity, but a completely legitimate C statement that reads something like, "add one to `x`". It goes a little deeper than this, but not much. A computer's memory is made up of many *words*, and with most machines produced in the 1990s each word is generally 32 *bits* long. A bit is a *b*inary dig*it* taking one of two values, 0 or 1, or, "on" or "off". Why? Computers are electrical beasts and an electrical switch is either on or off, giving two possible states. A voltage across two points is either zero or nonzero. Hence bits. Each byte (8 bits) of these words is given an *address*, or a unique location, in the computer's memory. For example, the *variable* `x` is stored in several bytes of memory and identified by the specific location of its first byte. What the above statement does is increment the number stored in that location by one. The `x` on the right-hand side represented the old value, `+` represents an operation performed by the computer's *central processing unit* (CPU), `1` is the second entity involved in the operation, and finally, the `x` on the left-hand side represents the memory address where the CPU puts the result.

Note the semicolon at the end of the above line of C code. It is very, very important. C is particularly touchy about programmers forgetting semicolons.

1.5.2 The `if` Statement

The `if` statement allows for conditional execution of blocks of code. In other words, suppose if a certain condition is true one block of code should be executed, but if the condition is false then a different block of code should be executed. We'll get into specific examples of why conditional execution is useful, but a trivial example is the statement embodied in the "flowchart" shown in figure 1.5.

Translating the flowchart concepts into C code gives

```
if(x<10)
{
        x = x + 1;

}
```

which is remarkably clear: If `x` is less than `10` then add `1` to `x`. We can extend the `if` statement, allowing different things to occur under different conditions. Suppose we have a sequence of two `if` statements

```
if(x<10)
{
        x = x + 1;
}
if(x>=10)
{
        x = x + 2;
}
```

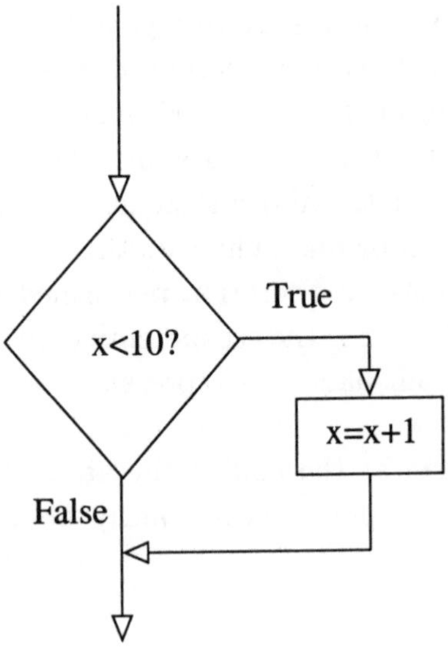

Figure 1.5. Flowchart of the statement `if(x<10) x=x+1;`. Program execution proceeds downwards. All allowable tests in an `if()` statement result in either true or false answers; if the answer is true, then additional lines of code are executed.

Again, the sequence of events is clear. If the initial value of `x` is 8, the final value is 9; if the initial value is 10, the final value is 12; and if the initial value is 9, the final value is 12. Most programming languages provide a replacement of these two sequential `if` statements with

```
if(x<10)
{
        x = x + 1;
}
else
{
        x = x + 2;
}
```

which gives a different final result of 10 if `x` initially takes the value 9. The above two examples, and their different outcomes, demonstrate the progress, or flow, of time as the code is followed downwards. Just as an aside, the above braces are unnecessary when the compound statement contained

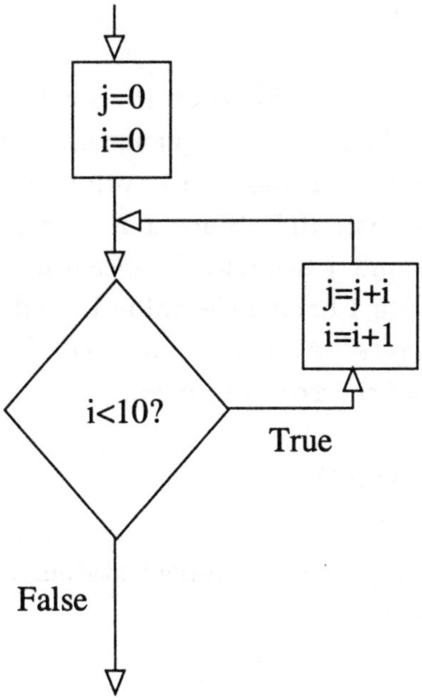

Figure 1.6. Flowchart of the `while` loop. Its most important feature is that program control passes back to the `while` test after execution of the lines following a true result. The `while` and `for` loops are effectively equivalent programming concepts.

within the braces consists of a single statement – we could therefore write more compactly[12]

```
if(x<10) x=x+1; else x=x+2;
```

1.5.3 The `while` *Loop*

In many ways the `while` loop is an extension of the `if` statement – the `while` loop allows a sequence of events to occur repeatedly as long as a specific condition is satisfied. For example

```
i = 0;
j = 0;
while(i<10)
```

[12] Although the meaning can change if the statement is surrounded by other `if`s. Use braces liberally.

```
{
        i = i + 1;
        j = j + i;
}
```

Let's construct the flow of this code (figure 1.6). First `i` and `j` are set to zero. Since `i` is less than 10, the sequence of statements is performed. After the end of the first pass, `i` and `j` are both one. The next pass begins with the test, "Is `i` less than 10?" Since `i` is 1, control passes inside the loop where `i` becomes 2 and `j` becomes 3. When all is done, control passes this looping structure with `i` having the value 10 and `j` having the value 55. *Can you check these results?* An equivalent, and shorter, representation of this code is written using the `for` statement

```
j = 0;
for(i=0;i<10;i=i+1)
        j = j + i;
```

the only benefit being a more compact statement. You could make it even more compact by writing

```
for(i=0,j=0;i<10;i=i+1) j=j+i;
```

or even

```
for(i=j=0;i<10;i++) j+=i;
```

(Note the shorthand notation `++` for incrementing by one[13] and `+=` for adding something to a variable.) Now you've got the real basics of any computer language, and in particular, C. All other features of a programming language are details, albeit important details like input and output statements, but most of the utility for scientific thinking is embodied in these simple statements.

1.6 A Few Words about Compilers

The programming statements described in the previous section, with other necessary statements all in the correct order, constitute a computer program. In actuality, they are part of a text file readable and editable by a human that knows how to program in C. The C programming language is one example of a high-level computer language. It's called "high-level" because the language means absolutely nothing to the computer and the instruction set contained on the computer's CPU. The CPU has a limited

[13] Hence, the nerdly origins for the naming of another programming language – C++ is an increment to C.

set of instructions directing the computer to fetch and return variables, add, subtract, and other basic operations. The CPU's language, however, is too cumbersome for programming because of details having nothing to do with the overall goals of the programmer.[14] A high-level language such as C allows us to ignore such details through the use of intermediate-level computer programs developed to serve as translators between the computer's CPU and humans. These computer programs are called compilers, and each programming language has its own separate and noninterchangeable compiler. Further, there are many compilers for each language, depending on both the CPU (e.g., Intel, Motorola). Some compilers are better than other compilers in terms of the efficiency of the machine-level code that is produced, and some compilers have lots of interactive features that make programming easier.

At present the two main computer languages for scientific computing are Fortran and C. Fortran has the well-deserved reputation of being a scientifically and mathematically oriented language. This reputation is built upon many years of use in a large variety of scientific fields. As a result, many functions scientists often need are built into the language and many other more specialized functions are contained in libraries accessible by computer programs. There are at least two versions of Fortran – F77 and F90 – named for the years that major revisions to the language were made. More recently, C has been developed and used as the basis of the Unix operating system, which makes it extremely compatible with the procedures controlling the computer. C++ is a recent elaboration of C that includes object oriented programming features, although the argument can be made that the additions are overly complicated and cumbersome for use in scientific programming. In addition, the Fortran function libraries can be used with C programs, allowing C programs to build upon decades of earlier work.

This book will discuss C exclusively. C was invented in the early 1970s at Bell Laboratories by Dennis Ritchie and grew out of a language named "Basic Combined Programming Language," or BCPL for short. Ritchie further shortened the name of his language to B. When the language was updated, it was called C. Presently there is a standard language called ANSI C, decided upon by a committee, but compilers also often work with an earlier version by Dennis Ritchie and Brian Kernighan (RK C). However, anyone can write whatever compiler they want, add extensions to the language, then sell it or give it away. Be cautious of nonstandard versions because the programs written for them may not work on other computers. People, computers, and

[14] If you are interested in learning more about the true essence of programming, seek out manuals on assembly languages.

compilers are constantly moving and changing: You don't want to spend a year developing code that will not work when you buy a new computer or upgrade your operating system or compiler. Hence make sure you learn to program according to the standards (at least as closely as possible). There are many references for C books. My first and favorite C manual is *The Spirit of C: An introduction to modern programming*, by H. Mullish and H. Cooper (1987). This favoritism dates me – the book predates the ANSI C standard adopted in 1990. Another good book, also predating the standard, is by Kernighan and Ritchie (1988). Newer books include Kelley and Pohl (1998), Oualline (1993), and Gottfried (1996).

1.7 The Personal Side of Programming

In the end, a computer program is a set of instructions a programmer sends to the computer. The computer (rather, the compiler) cares not one whit how the program looks to a human, nor the thought processes that went into its production. Even so, you might hear from other programmers that your code should look one way or another, or you should think this way or that when writing your programs. I want to spend a little time here dispelling some myths about programming.

Flowcharts When I learned to program back in the 1970s there was a great emphasis on using a flowchart to outline a program before putting it on the computer. On one hand, this process is a good one – think before you do too much. But really, the reason behind this emphasis was that back then programmers had to type their program on to punched cards then submit their programs to a computer operator who fed their program into the computer. The computer either ran the program if everything was done correctly, or it spat out error messages. If the latter, then the programmer had to figure out what went wrong (after griping at the computer operator), add, subtract, or modify cards, then submit the program once again. This whole process took hours of elapsed human time. It was worthwhile to spend more time up front making sure things worked. However, with the advent of personal computers, programmers edit their program files in a window on the screen and compiling the program is trivial. Writing a program is now an interactive process with the computer. Nonetheless, in scientific programming thinking can still be a useful tool – flowcharts are one way of putting your thoughts on paper, and subsequently into C code.

Comments You will hear from many sides, "Comment, comment, comment!" Comments are wise words of wisdom scattered throughout a program, but stripped out and ignored by the compiler. Comments are for the sole benefit of a human reading a program, and are especially important when a program is built by a committee – comments serve as communication between programmers. However, when a program is produced and run by a single person, as in the scientific programming I cover in this book, comments serve primarily as a reminder to oneself. The difference is important. I place relatively few comments in my code, preferring instead to make code as directly readable as possible using white space, functions, and well-chosen variable names. I usually add comments only in places where the algorithms get really detailed. But, to be honest, when I look at a program I wrote several years earlier, I often regret not having included more comments!

White space White space includes blank characters (spaces), tabs, and blank lines – all the empty space ignored by the compiler (excluding the necessary space between variable names and such). I use lots of white space. White space helps organize code into cohesive blocks and functional units, and this organization helps reduce the need for commenting. Of course, since it is ignored by the compiler, you don't need to add white space. If you want to maximize the number of programming statements viewable on your screen, white space simply displaces more statements.

Variable names At one point in time, variable names in computer programs were restricted to a letter followed by a few numbers or letters, so their names tended to be very cryptic. Using these variables was hard because you had to remember what each one represented, hence extensive comments were essential. Presently, variables can be almost any length (although be warned that some C compilers only look at the first eight characters) and include separating characters. These variables lead to very descriptive statements: `baby_prey = prey_reproduction_rate * adult_prey`. There is no need for a comment after such a line (unless for some reason the variable `adult_prey` represents the number of predators and the variable `prey_reproduction_rate` represents the predator mortality rate [the compiler does not care]). I make extensive use of descriptive variable names.

Functions A function (related entities include subroutines [Fortran] and procedures [Pascal]) is a collection of statements given a shorthand, descriptive name. Big blocks of code can then be called by another function by using a simple one-line statement. Functions enhance program readability manyfold by replacing blocks of code by descriptive functional identifiers. As with the enhanced clarity generated by long variable names, there would be no mistaking what happens if you call the function, `PreyReproduction()`. You need not place comments after such function calls (unless what really goes on in `PreyReproduction()` is predator death!). I often add function calls if either a block of code gets too big, or I need to repeat the same block of code in more than one place in a program.

2
Immigration–Emigration Models

Programming components covered in this chapter: • Essential UNIX and `vi` editor commands • Precompiler parameter definitions • `drand48()` random number generator • the `for` loop for repeated calculations • `if` statements for conditional tests • `printf` output statements • The `break` statement. • Using precompiler conditions

What are the processes that cause population numbers to rise and fall? The answer depends on at least one basic feature of the assumed ecological system – whether it is an open or closed system. A closed ecological system is one that is cut off entirely from the rest of the world, and population change is affected solely by the births and deaths of individuals within the closed system. In contrast, population change in an open ecological system is additionally affected by the immigration of organisms into the system and the emigration of organisms out of the system. In this chapter I cover open systems dominated by immigration and emigration, and in chapter 3 I address birth–death models of a closed system, forming the basis of more extensive stochastic models involving detailed biological questions (Bartlett 1960, Goel and Richter-Dyn 1974, Mangel and Tier 1993). My focus on a simple system in this chapter allows for the introduction of basic simulation and analytic models, as well as a chance to cover the essential mechanics of putting a computer program into the computer, compiling it, and running it.

Programs in these two chapters do not keep track of explicit individuals, only the total number of individuals in the simulated systems, keeping the C programs as simple as possible. In later chapters, after examining more advanced features of C, we will examine simulation models with explicitly represented individuals.

2.1 Immigration and Emigration Processes

The simplest model of an open system has individuals that come and go with constant immigration and emigration rates. A constant immigration rate means that over some short time interval there is a finite probability for one new individual to be added to the current population.

Emigration is slightly more complicated. Consider a population of birds on a small patch of land. At any time any bird might take off and leave the system, meaning that the emigration rate depends on the total number of birds in the population. This concept of emigration is identical to background mortality, and is a completely reasonable view of an ecologically motivated emigration process. However, another less ecologically relevant but pedagogically simpler emigration process is embodied by queueing theory (Karlin and Taylor 1975, 1981 – authoritative and mathematically rigorous). An example of a queue is a line in a grocery store governed by a checkout-clerk processing customers, taking the same time per customer no matter how many people are in line (figure 2.1A). Realistic ecological scenarios of organisms, other than humans, that emigrate in a queue-like manner may be rare. Consider, for example, a stream system with a pool that has a narrow inlet and outlet such that only one mayfly larvae can enter or leave during a short interval. This system is not an example of the simple queueing problem examined in this chapter because each individual in the pool has equal probability to leave during a given time interval (figure 2.1B). However, if the larvae take a number and forage in the pool until their number is called to move downstream, then the system becomes a simple queueing problem.

Independent of its ecological utility, a model containing these immigration–emigration processes provides a useful baseline examination of a stochastic system. In summary, the state of the system is completely defined by the number of individuals present at any time, and there are only two rules that govern the dynamics of the system:

(i) An individual immigrates to the system with rate α per unit time.
(ii) An individual emigrates from the system with rate β per unit time.

Deterministically, if the immigration rate α is greater than the emigration rate β, the population number grows linearly with time. If the emigration

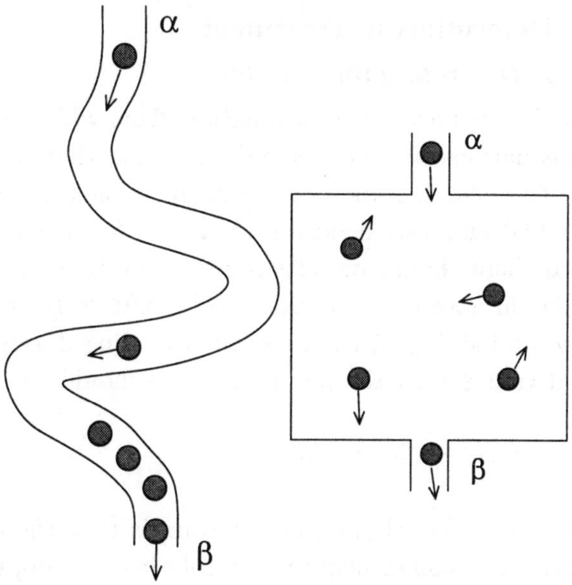

Figure 2.1. Alternative immigration–emigration models. Both panels depict a constant immigration rate α. (A) Queueing organisms. Their population's net emigration rate is independent of the queue size when the queue is greater than zero. (B) Non-queueing organisms. Their population's net emigration rate depends on the population number. This chapter examines the queueing model.

rate is larger than the immigration rate, then an initial population will decrease and hit zero population and stay at zero because populations must be nonnegative.

Stochastically, even when the population is expected to decrease, occasionally the numbers grow for a while, much like traffic on a freeway or customers through a grocery line. Also, once the line hits zero, immigration still occurs and the line can begin anew.

Queuing models have a rich history and are particularly useful because of their relatively simple solutions. The consequences of stochasticity can be examined and an understanding of the effects of various assumptions can be gained using these rather biologically unrealistic processes. In addition to stochastic models representing the dynamics of individuals, stochastic models can be built at the biologically higher level of groups, representing groups merging and splitting (Holgate 1967, Morgan 1976).

2.2 Deterministic Treatment

2.2.1 Discrete-Time Model

We begin with a discrete-time formulation, from which the continuous-time differential equations will be derived. Assume that at time t there is a population of size N_t. During a short time increment Δt, immigration occurs at a rate α and emigration occurs at a rate β. Thus, at the end of the time interval there should be $\alpha\Delta t$ immigrants in the population and $\beta\Delta t$ emigrants should have left. As long as $\alpha\Delta t, \beta\Delta t \ll 1$, we can think of these terms as the probability of these events occurring during the time interval Δt. Thus, at time $t + \Delta t$ the population size should be

$$N_{t+\Delta t} = N_t + \alpha\Delta t - \beta\Delta t. \tag{2.1}$$

The only major caveat about this equation is that the equation is invalid if the population becomes negative, simply because populations cannot be negative!

2.2.2 Continuous-Time Model

Using the definition of the derivative, we can obtain the continuous-time model from the discrete-time model. Rearranging (2.1) gives

$$\frac{N_{t+\Delta t} - N_t}{\Delta t} = \alpha - \beta. \tag{2.2}$$

The derivative from differential calculus is defined by shrinking the time step Δt to zero

$$\lim_{\Delta t \to 0} \frac{N_{t+\Delta t} - N_t}{\Delta t} \equiv \frac{dN}{dt} = \alpha - \beta, \tag{2.3}$$

where "$\equiv$" means "is defined by."

Both the discrete- and continuous-time model formulations imply three possibilities. First, if the emigration rate β exceeds the immigration rate α, then population numbers decline ($dN/dt < 0$). Second, if the immigration rate exceeds the emigration rate, then population numbers increase with time ($dN/dt > 0$). Finally, if the two rates exactly balance, the initial population density is maintained. These results simply confirm intuitive expectations.

2.3 Immigration–Emigration Simulation Model

2.3.1 Code

In this section I present a simulation of the immigration–emigration model discussed in section 2.1. This fully functioning program contains the essential parts of the C language. The name of the program, as it appears on the computer's file system, is `bd.c`. In the code listed below and elsewhere the numbers against the left-hand margin and followed by a right parenthesis are *not* part of the program, rather, these numbers are references for the following discussion.

```
1)    /* birth-death simulation */

2.a) #include  <stdlib.h>
2.b) #include  <stdio.h>

3)    #define   MAXTIME   100
4)    #define   PRNTTIME  10

5)    int main(void)
6)    {
7)         int  ttt, event, n;
8)         int  seed;
9)         double    alpha, beta;

10)        alpha = beta = 0.1;
11)        n = 10;

12)        seed = 1456739853;
13)        srand48(seed);
14)        printf(" time   pop\n");

15)        for(ttt=0;ttt<MAXTIME;ttt++)
16)        {
17)             event = 0;
18)             if(drand48()<alpha) event = 1;
19)             if(drand48()<beta) event = event - 1;
20)             n += event;
21)             if(n<0) n = 0;

22)             if(ttt%PRNTTIME==0)
                     printf(" %3d   %-4d\n",ttt, n);
23)        }
24)        return (0);
25)   }
```

This program represents the kernel of future programs, so it will be important to understand its workings. If programming is new to you, it will be helpful to go through this simple code line-by-line. If you are familiar with other languages, but new to C, look for syntax that is different from your other languages. The code is written in the "new" style of C for Unix machines. At my web page are versions written in the "old" style of C for Unix machines (both versions compile with the same compilers) and a new style version for the LCC-Win32 compiler for PCs.[1]

One of the most important aspects of programming is readability. I use white space generously, indenting lines and adding blank lines between related statements. Both of these things are completely ignored by the compiler, so you're free to develop your own style.

2.3.2 Code Details

Line 1: Everything inclusive of `/*` and `*/` represents a comment and is ignored by the C compiler.

Line 2: Any line beginning with a `#` symbol represents a command processed by the precompiler. The precompiler is an extremely useful part of the C compiler that adds additional files, picks out specific parts of code, and expands any shorthand notations that the programmer used throughout the program. It is called automatically whenever the compiler itself is called, and does not require any additional processing on the user's part. These lines tell the precompiler to include the `stdlib.h` and `stdio.h` header files, which contain information needed for calls to standard C library functions. Here `stdlib.h` is included to call the random number generator `drand48()` (see lines 12–13 and 18) and `stdio.h` is included to use `printf()` (see line 22).[2]

Lines 3 and 4: These lines are precompiler definitions. The precompiler replaces all subsequent occurrences of `MAXTIME` and `PRNTTIME` with `1000` and `10`, respectively. These definitions may seem useless, but the great advantage is that if, for example, `MAXTIME` occurs in a dozen places throughout the code and you want to change its value, replacing it once in the definition statement and letting the precompiler replace it everywhere else is easier than searching and replacing it everywhere by hand.

[1] The primary difference between the PC and Unix versions is the use of `drand48()` random number generation – the PC version should compile on Unix machines.

[2] With some compilers these `include`s are not needed to invoke these particular functions, but in any case, the inclusions will not disrupt anything.

Line 5: Every program must have an `int main(void)` procedure that represents the first procedure invoked by the operating system when a program is run. I always place my `main` procedure as the first procedure in the program file. The `int` in front of the function name means that the operating system expects the program to return (see line 24) a particular type of number (see lines 7–9) after the program finishes execution. Within the parentheses is the word `void` denoting that the program expects nothing to be passed to it from the operating system when execution starts.[3]

Line 6: An open brace, `{`, represents the beginning of a series of statements that taken together are called a compound statement. The compound statement ends with a matching closing brace, `}`, in this case shown on line 25.

Lines 7–9: All variables used within a procedure must be defined somewhere. In this case the variables are defined here at the beginning of the procedure. These three lines define two different types of variables, `int` and `double`. `int` variables use a single 32-bit computer word to represent an integer number. Thus the numbers that can be represented range from -2147483648 (-2^{31}) to $+2147483647$ ($2^{31}-1$). There are actually (at least) two types of `int` variables, `short int` (2 bytes) and `long int` (4 bytes). I use `int` synonymously with `long int` because my compilers use it as the default for `int` variables. Variables of type `double` use two computer words, but divide the bits into two parts representing separately a mantissa and an exponent. Imagine a number written in scientific notation, $\pm$x.xxxx $\times 10^{\mathrm{y}}$. The x.xxxx is the mantissa and the y is the exponent. Variables of type `float` are similar but use only one word. `floats` have only about 6 digits of accuracy ranging between $\pm 3.4 \times 10^{\pm 38}$, `doubles` have about 15 digits of accuracy ranging between $\pm 1.7 \times 10^{\pm 308}$. `doubles` have more accuracy, but `floats` are faster. I always use `doubles`.

Lines 10–11: These are the first real statements that do anything, at least in terms of the program's logic. What went before were just the preliminaries. These are simple assignment statements, discussed in the Introduction, that assign specific values to the three variables listed.

Lines 12–13: These two lines initialize the random number generator (technically and importantly, pseudo-random number generator). All

---

[3] Later on we examine command line arguments that pass parameter values to your program (see page 201), in which case we replace the `void` argument with nontrivial arguments.

streams, or lists, of random numbers begin with a *seed*. The generator used in this program uses a seed of type `int`. Here the seed is explicitly given, and every time the program is run, the identical stream of "random" numbers will be produced, unless the seed is changed. The seed is passed to the system via the `srand48()` call. This call is a bit of a magic box, but you can gain much information on it by examining the output of the `man srand48` Unix command. Be very careful of random number generators, especially the simple ones that you might find written down in books. There can exist correlations between successive numbers, or every other number, or every third number, ... , or collections of three or more numbers. As the saying goes, "Random numbers are far too important to be left to chance." In chapter 4 we'll examine random number generation in great detail.

Lines 14: This line prints out a heading for the columns of data printed out when the program runs. Everything within the pair of double quotes is output to the display, except that the `\n`, one of the more commonly used control sequences, represents a carriage return and line feed in the output.

Lines 15 and 16–23: This looping statement repeats (or iterates) the block of statements (a compound statement) contained within the opening and closing braces on lines 16 and 23. At the start of the loop, the variable `ttt` is set to 0. Next, a test is performed asking whether `ttt` is less than `MAXTIME`. If the test turns out to be true, then the compound statement is executed. After execution, the variable `ttt` is incremented by one (`ttt++` reads `ttt=ttt+1`), and control returns to the point where `ttt` is tested against `MAXTIME`. When the test is false, program control is passed beyond the closing brace on line 23.

Line 17: The variable `event` is set to zero. If this were not done, its value from previous iterations would be carried over to the present iteration.

Line 18: `drand48()` is a C library function that returns random numbers. We will use it often in the programs ahead. After initializing the function through the `srand48()` call on line 14, subsequent calls to `drand48()` return a number distributed evenly between 0.0 and 1.0 (0.0 can be returned, but not 1.0).[4] You can think of the function just like a number.

[4] PC Users: If these two functions fail, try replacing `srand48(seed)` with `srand(seed)`, and `drand48()` with `(double)rand()/((double)RAND_MAX+1)`. These random number generation functions should be available on all platforms, but use `drand48()` if available.

If the random number is less than the birth probability, **alpha**, then **event** is set to 1, representing an immigration event. Hence, if **alpha** is 0.1, 10% of the tests result in births. If the random number is greater than **alpha**, control passes to the next statement.

Line 19: This line is just like line 18, except it tests for a death and decrements **event** if a death occurs.

Line 20: The value contained in **event** is added to the total population size (**n+=event** reads **n=n+event**).

Line 21: It is possible for the population size to go below zero (if a death occurs when the population is zero), which makes no biological sense. This line resets the value to zero.

Line 22: The time and population size every **PRNTTIME** simulation steps are printed by this line. The **%** in the expression **ttt%PRNTTIME** is called the "modulus" operator which gives the remainder after having divided **ttt** by the constant **PRNTTIME**. For example, 7%4=3, 8%4=0, and 9%4=1. Also note the double-equal signs, **==**, used inside the parentheses. These indicate a test of equality rather than an assignment statement. Always use **==** for the tests of **if** statements. Just to make life more difficult, not all instances of **%** mean the same thing. In the **printf** statement, the expression **%3d** represents the field used to output the variable **ttt**, and tells the compiler to print (or display) the value right-justified in a field of three spaces. The next variable, **n**, is printed in a left-justified field (the minus sign specifies left-justification) of 4 spaces. Another important output format is **%f**, used to print variables of type **float** or **double**.

Line 23: This closing brace ends the compound statement, begun on line 16, associated with the **for** loop on line 15.

Line 24: Upon finishing execution of the program, **main** is supposed to pass an **int** value back to the operating system. The **return** statement accomplishes this feat. Although nothing useful happens here, later on we'll use **return** to pass values back from functions (see page 58).

Line 25: This closing brace ends the **main()** function begun on line 6. All the statements between lines 6 and 24 represent a single compound statement.

2.4 Getting the Program into the Computer

There are two equally important aspects to putting the above code into the computer allowing you to then compile and run it. First you must be able to use the operating system well enough to start a program that edits text files. Second, once you start the editor, you clearly need to know how to use it. This section covers both these topics for Unix computers. If your machine does not use Unix, then you should skip this section and figure out how to enter your `bd.c` program such that it is readable by your compiler. Then turn to page 36, section 2.5, where we compile the program and run the executable.

2.4.1 Basic Unix Commands

You only need a small set of Unix commands to get through life. Realize that there are an infinite number of Unix commands – Unix is just one big C program and the little C programs scientists write are user-defined Unix commands. BEWARE: You can lose your soul playing with Unix and it may be useful to put on blinders and refuse to learn more details.

After logging into your account on a Unix machine, usually there is a command tool (or window) available that accepts Unix commands. Also, this command tool should have your home directory as its working directory. All of the commands listed below are commands typed in this command window.

The command `ls` or `ls -a` or `ls -l` lists all the files you have in the directory of your account. The `-a` flag lists the hidden files (those starting with a `.`) along with the others. The `-l` flag displays various information about the files such as permissions, creation dates, and sizes. Try typing

```
ls -a
```

The `a` stands for "all files", and if you've never used your account there may be nothing but files starting with a period "`.`". Try typing

```
ls
```

and see what difference the flag makes.

At this point make a new directory to hold your first C program. I really like setting up and organizing a directory for each project I work on to store the associated program and data files. Much like the folders in filing cabinets or Macintosh and Windows operating systems, directories provide this organization. To create a directory, type

Table 2.1. Essential Unix commands

`ls`	list files in current directory
`ls -a`	include hidden files in directory list
`ls -l`	show file details in directory list
`mkdir` *dirname*	create a new subdirectory named *dirname*
`cd` *dirname*	change current directory to *dirname*
`cd ..`	move to parent directory of current directory
`more` *fname*	show contents of file *fname*
`mv` *fname1 fname2*	move file *fname1* to *fname2*
`cp` *fname1 fname2*	copy file *fname1* to *fname2*
`man` *command*	display information on Unix commands

```
mkdir brthdth
```

and hit return. The first word, `mkdir`, is a Unix command that makes a new directory, in this case called `brthdth` for the "birth–death" models that we will examine in these first few chapters. Now change "location" from the home directory to the `brthdth` directory by typing

```
cd brthdth
```

where `cd` stands for change directories. Try listing the contents of this directory by typing the command `ls`. Nothing should be returned because `brthdth` is a new directory. Now type

```
cd ..
```

then hit return. The two periods always refers to the parent directory above the present directory. If you list the contents of the present directory, you should see the new `brthdth` directory listed.

2.4.2 The `vi` Editor

This section introduces the fundamentals of editing a program file using a standard Unix editor called `vi`. This editor is very useful because it is available over any `telnet` (or the more secure `ssh`) session. Other editors with this capability include the very popular editor, `emacs`, and `pico`. Likewise, if you are sitting at a Unix machine, you still have `vi` available to you through a command window, but I find `vi` more difficult to work with than editors such as `textedit` or `nedit`. I will describe in detail only the few `vi` commands needed to enter the above code into a file. Table 2.2 provides a short list summarizing the important commands (after the man page on Sun).

First use `cd` to go into your `brthdth` directory. We want to begin a `vi` editing session:

- **opening your file (`vi filename`)** Type `vi bd.c` in your command window (followed by the ubiquitous "return") and you should see a screen full of tildes, ~, with a bottom line containing `"bd.c" [New file]`, and a cursor at the top left character position. If so, you're ready to enter the code. If not, you have a problem and you should contact your system administrator or other knowledgable person.

Now that you have opened the file, try closing it before doing anything else.

- **closing your file (`:q`)** Type the character : (by pressing the "Shift" and the `:/;` key simultaneously) to get a command line at the bottom of the `vi` screen. The line begins with a : and the cursor is placed after it. Now type `q` and press return. The `vi` screen will disappear. Since no edits were performed on the empty file nothing was saved, and a listing of the directory (`ls`) shows no evidence of the file `bd.c`.

Now reopen your file with `vi bd.c`. Here I take you through the basic `vi` commands enabling you to enter the entire program. Adding new text requires a new `vi` command:

- **inserting text (`i`)** Type the character `i` to enable the insertion of new text (in front of the character the cursor overlaps). No change will be observed on the screen. Now type `/* birth-deeth simulation */` and if you make any errors in addition to the misspelling of `death` just continue on to the end of the line. After typing the last backslash, press the "Esc" (escape) key on your keyboard. You should see your cursor back up one space to overlap with the last backslash. Anything you type now will be interpreted as a command, so watch out.

At this point the goal is to save your file and exit `vi`. This twofold task is accomplished by:

- **saving and closing your file (`:wq`)** Type the character : to get a command line, then type `w` to write (or save the changes) to the file in your directory, then type `q` and press return to exit `vi`. A listing of the directory (`ls`) should now show the file `bd.c`.

Table 2.2. Essential `vi` text editor commands. `CR`=`Return` key, `ESC`=`Esc` key, and `^`= `Control` key (held down like the `Shift` key). Upper case implies using the `Shift` key.

File commands	
:wqCR	exit vi; save changes
:q!CR	quit, discard changes
Search commands	
/*text*CR	search for *text*
f*x*	find next *x*
F*x*	find prev *x*
Editing commands	
i*text*ESC	insert *text*
x	delete a character
dw	delete a word
dd	delete a line
3dd	delete 3 lines
u	undo previous change
Backspace	erase last character
^W	erase last word
Cursor and screen changes	
← ↓ ↑ →	move the cursor
h j k l	same as arrow keys
^U ^D	scroll up or down
^F	forward screen
^B	backward screen
H	top line on screen
L	last line on screen
CR	next line at first non-white
0	beginning of line
$	end of line
w	forward a word
b	back a word
e	end of word

Reopen your file with `vi bd.c` to correct the spelling of `death`. The file is reopened with the cursor at the top left corner of the screen, overlapping with the first backslash. Place the cursor over the second `e`, requiring cursor movement commands:

- **moving the cursor (arrow keys: ← ↓ ↑ → or letter keys: `h` `j` `k` `l` respectively)** Type either the right arrow or the `l` key until the cursor overlaps the second `e` in `deeth`.

Correcting the misspelling requires another new `vi` command:

- **replacing a single character (`r`)** Press `r` for replace, then `a` for the replacement character. You will see the `e` change magically into an `a`.

Many more `vi` commands exist and can be discovered in many Unix texts or by typing `man vi` in the Unix command window (if you do the latter, press the space bar to read successive pages of the help file displayed in your command window).

2.5 Compiling the Birth–Death Program

I assume the program now resides somewhere on your computer.[5] Assuming you named the program `bd.c`, it is compiled on a Unix machine by typing the statement

```
cc bd.c -o xbd
```

and pressing the return key.[6] This command invokes the compiler to convert the human-readable text file `bd.c` into the computer-readable binary file `xbd` specified after the `-o` flag.[7] Without this latter specification, the program is compiled into the file `a.out`.

If compilation was successful, the computer replies with a normal prompt. If the command gave any error messages, perhaps the compiler statement above was typed incorrectly, you have a typo in the program (look for missing semicolons!), or something else went wrong. After making sure the first two possibilities are not the problem, try to find someone who already knows C. The program uses no complicated features of C – an experienced programmer will likely find the problem very quickly.

[5] Programs are stored on the hard drive; programs are run after being loaded from the hard drive into memory.

[6] If `cc` fails, you might try `gcc`, or even `CC`.

[7] A flag is a command line argument that indicates a desired option to a command (see page 201).

The program is executed by typing in **xbd** on the next command line, which should produce the output

```
xbd <return>
 time   pop
   0    9
  10    10
  20    8
  30    7
  40    6
  50    7
  60    9
  70    11
  80    12
  90    11
```

If you get numbers identical to the above results, everything should be working fine. If the numbers are different, but demonstrate similar temporal dynamics, perhaps your random number generator is different (e.g., the numbers will be different if you use **rand().**). If the results give constant values for n, there is a mistake somewhere in your program.

If you get the same results I did, try running your program one more time. Do you get the same numbers once again? You'd better! This repetition occurs because the same random number generator seed was used for both runs, and because the random number stream (the series of random numbers generated) is deterministically produced once given the seed. Try changing the random number seed, recompiling the program, then rerunning your executable file (in this case, **xbd**). This time the output should have changed.

2.6 Program Extensions

2.6.1 Code Changes

Making a few changes to the bd.c code yields some useful results beyond the above list of numbers. I implemented the following changes to produce the results described in the next section.

Averaging over multiple runs. Beneath line 4, add a new definition statement

```
#define   NUMRUNS   10
```

then add a loop outside the **ttt** loop that runs over an integer variable **nrun**

```
for(nrun=0;nrun<NUMRUNS;nrun++)
```

Make sure to define the new variable, add braces around the new loop's compound statement, and reinitialize **n** for each new run. Define a few new variables of type **double** to store the run averages taken at specific times during the simulations. For example, if **ave50** represents the average taken at time 50, just before line 22, add

```
if(ttt==50) ave50 += n;
```

(Be sure to set these averages to 0.0 before using them.) When all the runs have finished, divide the averages by **NUMRUNS**, then output the averages. Make sure the print statements are outside the **nrun** loop!

Alternative model assumptions. I enabled using two different model assumptions: (1) whenever a run hits zero population, that run stops, and zeroes are counted for the remainder of the time; or (2) runs that hit zero can have immigrants "reseed" their population counts. I accomplished this change by adding a switch (with a comment stating my options)

```
#define RUNTYPE 1 /* 1 - zero stays extinct,
                     2 - reseeding */
```

and replacing line 21 with

```
#if(RUNTYPE==1)
        if(n==0) break;
#endif
#if(RUNTYPE==2)
        if(n<0) n = 0;
#endif
```

These **#if** statements are extremely useful. The precompiler first runs through the program and, depending on the value of **RUNTYPE**, incorporates or excises portions of the code that are satisfied or not satisfied by the precompiler's conditional test of the parameter value. After the precompiler finishes, the compiler compiles what remains. In this way, by simply setting the value of **RUNTYPE**, two different programs can be generated.

2.6.2 Simulation Results

I ran the programs in a command tool window and, like the previous run, the program's **printf** statements output to this window. This window is the *standard output* "file," abbreviated by **stdout**; you can also use a slightly more complicated output function, the **fprintf()** statement to output directly to a file (see page 94). An alternative approach to getting the data into a file is to use Unix's redirection command to redirect the output from `stdout` to a text file. To make this data file, just rerun the program, but this time use the command

```
xbd > datafile
```

Once the program finishes, list the files in your directory – a new entry for `datafile` will be there. You can examine its contents using `more datafile`. Given the data file, I used a Unix plotting program, `xmgr`, to produce the graphs, but a spreadsheet program could be used to import these text files and plot the data.

Run stochasticity. Figure 2.2A shows results from ten different runs of the immigration–emigration program with both immigration and emigration rates equal to 0.1 per unit time. Each run could either use a different seed for ten different program runs, or use one seed with each run performed sequentially within a single execution of the program. Either way, a different *stream* of random numbers is used for each of the ten runs. I used a starting population of ten individuals for each run. Notice that many of the runs go extinct and, once extinct, never get another immigrant. You could certainly argue about the ecological appropriateness of such model assumptions. If these rules are indeed depicting an open system with immigration and emigration, immigrants should not be prevented from arriving at an empty system. Likewise, if this is a model of a closed system, the number of births and deaths should depend on the number of individuals present at any time. The biological problems implicit in these interpretations will be alleviated in the birth–death models of chapter 3.

Run averaging. Figure 2.2B shows the results of averaging over a number N of independent runs. The extensions discussed in section 2.6 represent one possible way of obtaining these data (N and `NUMRUNS` are the same quantity). A single run ($N = 1$) has a great deal of variation, much of which is removed by averaging over as few as $N = 10$ runs. Averaging over $N = 10000$ runs gives a rather flat line, in congruence with the deterministic model predictions for equal immigration and emigration rates, $\alpha = \beta$. Conceptually, this flatness is reasonable because immigration should exactly balance emigration when the rates are equal. Using *arrays* to hold lots of data, introduced in chapter 3, greatly simplifies computing these averages. As I described in section 2.6, there are ways to avoid arrays, but when the amount of data is very large (as in these plots) those methods are extremely inefficient.

Immigration to extinct systems. Figures 2.2C,D demonstrate the importance of one seemingly small assumption regarding what to do with extinct systems. Figure 2.2 assumed that extinct systems (individual counts

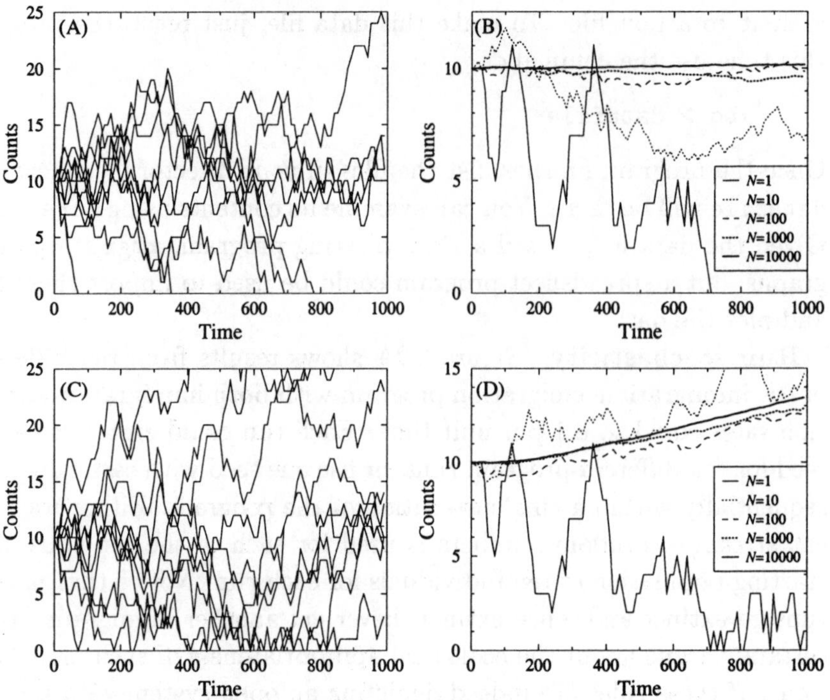

Figure 2.2. Results when extinction is (A,B) permanent (RUNTYPE=1) and (C,D) not permanent (RUNTYPE=2): (A,C) Ten runs using different random number streams. (B,D) Results from averaging over sets of N runs.

of zero) remain extinct forever, but here extinct systems can be reseeded by immigrants. This reseeding is perfectly reasonable – why should immigrants avoid empty systems? With reseeding, the averages tend to increase, contrary to the intuition that immigration exactly balances emigration when the rates are equal.

In the next section I address two issues using an analytic stochastic model. First, why do the averages in figure 2.2D increase, whereas the deterministic model predicts constancy? Second, the deterministic analytic model addresses the average of many runs, but has no predictions for the variance in a collection of runs.

2.7 Analytic Stochastic Immigration–Emigration Model

2.7.1 Mathematical Analysis

The analytic model discussed in this section is important for two reasons. First, it demonstrates that stochastic simulation models indeed have analytic counterparts that capture the stochasticity, and, second, even the simplest of these analytic formulations can be rather formidable. As mentioned above, the general modeling problem is within the framework of queueing theory, or the theory of "queues," describing common problems in human behavioral ecology – for example, the dynamics of grocery checkout lines and the number of cars on roads.

Karlin and Taylor (1975, 1981) (THE authority on stochastic processes) discuss a model that describes the individualistic nature of the transitions between states.[8] The question is posed as finding the probabilities for the system to be in particular states, with each state defined by the number of individuals. Hence $P_0(t)$ represents the probability for the system to have no individuals at time t. More generally, $P_n(t)$ defines the probability to have n individuals at time t. Another way to think about these probabilities is to imagine an enormous number of independent systems – then $P_n(t)$ represents the fraction of systems having n individuals at time t.

Assuming that the time step Δt is small enough such that only transitions of $\Delta n = \pm 1$ need be considered, then the probability densities are coupled by the set of relationships

$$P_n(t + \Delta t) = P_n(t) + \alpha \Delta t P_{n-1}(t) + \beta \Delta t P_{n+1}(t) - (\alpha + \beta)\Delta t P_n(t). \tag{2.4}$$

Thinking about P_n as the fraction of many independent systems having n individuals, the first term on the right-hand side of the equal sign represents the fraction of systems with n individuals at the previous time t. All of the remaining terms are represented by one of the arrows in figure 2.3. The second term represents systems having $n - 1$ individuals gaining an additional individual through immigration. This additional fraction is a product of two terms, the probability of an immigration event happening during a time Δt, $\alpha \Delta t$, and the fraction of systems previously having $n - 1$ individuals. The next term, involving emigration with rate β from systems having $n + 1$ individuals, has an analogous interpretation. The last term accounts for the immigration or emigration of one individual from systems previously having n individuals.

Rearranging (2.4)

$$\frac{P_n(t + \Delta t) - P_n(t)}{\Delta t} = \alpha P_{n-1}(t) + \beta P_{n+1}(t) - (\alpha + \beta) P_n(t) \tag{2.5}$$

and taking the continuous-time limit, $\Delta t \to dt$, yields the set of ordinary

[8] Excellent treatments of biologically relevant stochastic processes are presented by Bartlett (1960), Goel and Richter-Dyn (1974), Nisbet and Gurney (1982), and Renshaw (1991).

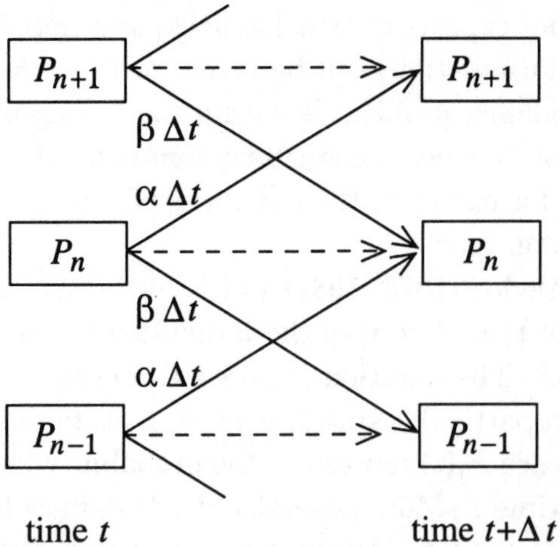

Figure 2.3. Graphical representation of the immigration–emigration probability density dynamics. Given a system with n individuals during the time interval Δt, an immigration occurs with probability $\alpha\Delta t$, taking the system to size $n+1$, and an emigration occurs with probability $\beta\Delta t$, taking the system to size $n-1$.

differential equations

$$\frac{dP_n(t)}{dt} = \alpha P_{n-1}(t) + \beta P_{n+1}(t) - (\alpha + \beta)P_n(t). \tag{2.6}$$

The equation for the empty probability, $P_0(t)$, is a special case

$$\frac{dP_0(t)}{dt} = \beta P_1(t) - \alpha P_0(t). \tag{2.7}$$

The model outlined above assumes that extinct systems can be restarted, as is the case in figure 2.2C,D, but not figure 2.2A,B.

We are not going to examine the full dynamics of the system. We will solve only for the equilibrium probability distribution, or, in other terms, the set of steady-state probabilities $P_n \equiv P_n(t \to \infty)$. Given steady-state conditions, the above set of dynamical equations becomes

$$0 = \beta P_1 - \alpha P_0 \tag{2.8}$$
$$0 = \alpha P_{n-1} + \beta P_{n+1} - (\alpha + \beta)P_n. \tag{2.9}$$

The first of these equations leads to

$$P_1 = \frac{\alpha}{\beta} P_0. \tag{2.10}$$

The next equation in the series, equation (2.9) with $n = 1$, is

$$0 = \alpha P_0 + \beta P_2 - (\alpha + \beta) P_1, \tag{2.11}$$

and yields

$$P_2 = \left(\frac{\alpha}{\beta}\right)^2 P_0. \tag{2.12}$$

A little bit of further manipulation (see exercise 2.6) shows that

$$P_n = \left(\frac{\alpha}{\beta}\right)^n P_0. \tag{2.13}$$

Equation (2.13), the probability distribution of the immigration–emigration problem, is called a geometric distribution. Requiring that the sum of all probabilities equals 1 leads to the geometric series (if $\alpha < \beta$), $1+r+r^2+r^3+\cdots = 1/(1-r)$, where $r = \alpha/\beta$, and a value for $P_0 = (\beta - \alpha)/\beta$. Using the steady-state probability P_0, all others are obtained through equation (2.13). Given the resulting probability distribution, you can calculate the mean

$$m = 0 \cdot P_0 + 1 \cdot P_1 + 2 \cdot P_2 + \cdots = \sum_{n=0}^{\infty} n P_n, \tag{2.14}$$

and variance

$$\mathrm{Var}(n) = \sum_{n=0}^{\infty} (n - m)^2 P_n. \tag{2.15}$$

The mean and variance of the geometric distribution can be derived (see exercise 2.7) or found in either an elementary text on probability or a mathematical handbook, giving

$$m = \frac{\alpha}{\beta - \alpha} \tag{2.16}$$

$$\mathrm{Var}(n) = \frac{\alpha\beta}{(\beta - \alpha)^2}. \tag{2.17}$$

Note that the results are only sensible for $\alpha < \beta$, and when $\alpha = \beta$ the mean and variance diverge to infinity. Are these predictions in agreement with the simulations demonstrated above?

2.7.2 Comparison of Simulation and Analytic Models

There are now analytic predictions against which we can test simulation results. To make this test, I revised the simulation code to sweep through a series of values for the immigration probability α while measuring the

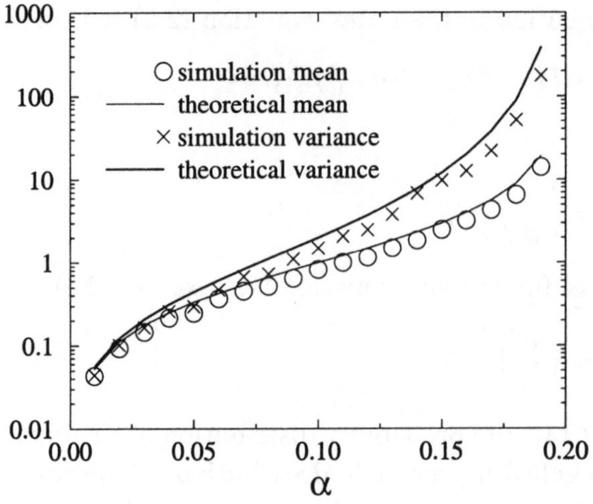

Figure 2.4. Comparison of results from long time averages from the simulation model (data points) with the analytic predictions (curves). Why such large discrepancies between the two model formulations?

population mean and variance (see exercises 2.1 and 2.4). This revised code produces the data displayed in figure 2.4. Here I used the model that assumes immigrants can arrive at an empty system to better match the assumptions of the stochastic immigration–emigration model of the previous section. I set $\beta = 0.2$ and the initial population to $N = 10$ for all runs, and I minimized the effects of initial transients by throwing away the first 500 simulation steps. After this initialization, I then took measurements every 100 simulation steps over an elapsed time of 200,000 simulation steps.

On the one hand, there is good qualitative agreement between the simulation results and the analytic predictions: Analytic predictions for both the mean and the variance nicely match the dependence on the immigration rate. On the other hand, the simulation values are consistently beneath the analytic curves. Why?

Let's look more carefully at the heart of the simulation program

```
event = 0;
if(drand48()<alpha) event = 1;
if(drand48()<beta) event = event - 1;
n += event;
```

What really happens here is that if the first random number is less than α *and* the second random number is greater than β, then an individual is added to the population. However, if the first random number is less than

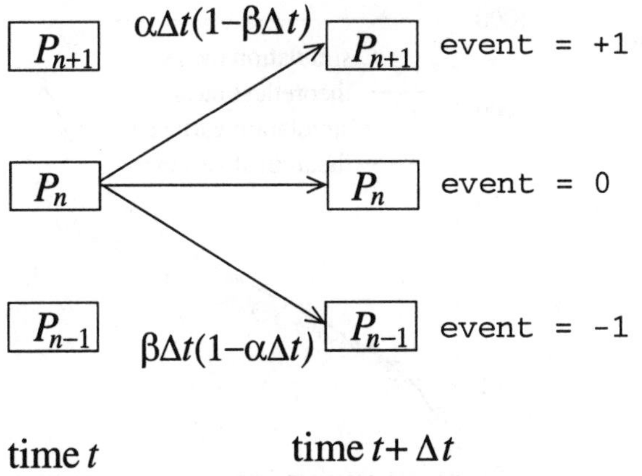

Figure 2.5. Graphical representation of the immigration–emigration rules as coded in `bd.c`. Compare these transition probabilities with those of figure 2.3.

α *and* the second random number is less than β, then no individuals are added. A similar difficulty arises when considering emigration. This cross-term between immigration and emigration mixes up the two, supposedly independent, processes and results in the probabilistic transitions depicted in figure 2.5 instead of the intended transitions depicted in figure 2.3.

However, the stochastic immigration–emigration model, equation (2.4) (reproduced here)

$$\begin{aligned} P_n(t+\Delta t) &= P_n(t) + \alpha \Delta t P_{n-1}(t) \\ &\quad + \beta \Delta t P_{n+1}(t) - (\alpha+\beta)\Delta t P_n(t)) \end{aligned}$$

has no cross-terms mixing up the two processes. A better analytic representation of the simulation is

$$\begin{aligned} P_n(t+\Delta t) &= P_n(t) + \alpha\Delta t(1-\beta\Delta t)\, P_{n-1}(t) \\ &\quad + \beta\Delta t(1-\alpha\Delta t)\, P_{n+1}(t) \\ &\quad - (\alpha(1-\beta\Delta t) + \beta(1-\alpha\Delta t))\Delta t\, P_n(t)) \end{aligned} \tag{2.18}$$

which can be mapped back to equation (2.4) with different parameters $\alpha' = \alpha(1-\beta\Delta t)$ and $\beta' = \beta(1-\alpha\Delta t)$. Figure 2.6 reexamines the simulation data by using α' as the independent variable and comparing the results to the mean (equation (2.16)) and variance (equation (2.17)) replacing the unprimed parameters with primed ones. Now the matches between simulation results and analytic predictions are rather good.

I tried to demonstrate two things in this chapter. First, simulations of stochastic processes have their counterparts in the mathematical world.

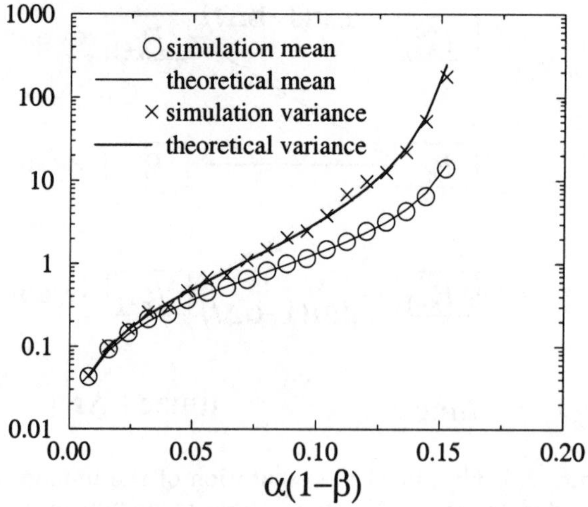

Figure 2.6. Comparison of simulation and analytic results accounting for the cross terms in birth and death processes (using $\Delta t = 1$). Simulation data are identical to the data of figure 2.4, but just replotted as shown. Excellent agreement is seen between the two model frameworks.

When the interactions are not too complicated, expressions can be derived for probability distributions that provide comparisons to simulation measurements. Second, predictions from analytic models can provide benchmarks against which to compare simulation results. A well-written simulation and an appropriate analytic model should match: Disagreement between two such models means that one or both models are wrong. Hence, an analytic model can serve to clarify and correct the logic in a simulation model.

2.8 Exercises

2.1 Incorporate the extensions from section 2.6. Print out for times 200, 400, 600, 800, and 1000, averages over 10, 100, 1000, and 10000 runs for `RUNTYPES` 1 and 2. Do your results match those of figures 2.2B and 2.2D?

2.2 Assuming that the immigration–emigration model exemplifies a grocery line, compare the fraction of time a checkout clerk has no people standing in line with the expected line size for a series of ratios β/α.

2.3 Do you need to worry about a limitation on eight-character variable names? Type in this small program, then compile and run it to find out.

```
int main(void)
{
    double variable_name_1, variable_name_2;
    variable_name_1 = 10.0;
    variable_name_2 = 20.0;
    printf("%f %f \n", variable_name_1, variable_name_2);
    return (0);
}
```

2.4 The mean and variance of a set of n numbers, x_1, x_2, x_3, ..., x_n, is defined as

$$\begin{aligned}\mu_n &= \frac{1}{n}\sum_{i=1}^{n} x_i \\ V_n &= \frac{1}{n}\sum_{i=1}^{n}(x_i - \mu_n)^2,\end{aligned}$$

however, this pair of equations is cumbersome to apply in practice because the variance calculation requires knowledge of the mean. Usually a computer program produces the list of numbers to average one at a time. Using the above variance equation requires storage of all the numbers produced until the mean is finally determined after the final number is produced. A better scheme updates the mean and variance determined from a list of $n-1$ numbers to include the new datum x_n using the three equations

$$\begin{aligned}\Delta &= x_n - \mu'_{n-1} \\ \mu'_n &= \mu'_{n-1} + \frac{\Delta}{n} \\ SS'_n &= SS'_{n-1} + \Delta \times (x_n - \mu'_n).\end{aligned}$$

Show mathematically that $\mu'_n = \mu_n$ and $SS'_n = nV_n$.

2.5 Show that equation (2.9) is consistent with the hypothesized probabilities $P_n = (\alpha/\beta)^n\, P_0$.

2.6 Derive equation (2.13). (Hint: First try deriving equation (2.12).)

2.7 Derive the mean (equation (2.16)) and variance (equation (2.17)) for the birth–death model. (Hint: $\Sigma n r^n = r\frac{d}{dr}\Sigma r^n$).

2.8 The right way to print an integer variable `i` and real variable `x` is, for example, `printf("%d %f \n",i,x);`. Test out a wrong way

```
int main(void)
{
    int i=12;
    double x=12.0;
    printf("int as double: %f \n",i);
    printf("double as int: %d \n",x);
    return (0);
}
```

2.9 Modify the birth–death simulation code to calculate the probability distribution P_n. You could use either a time average from a single run or an ensemble average from many different runs. Using the former, let the system come to equilibrium and then begin tracking the population counts. Using the latter, from an arbitrary number of runs track the population numbers at the end of the simulation. Calculate the mean and variance using the algorithm of exercise 2.4 and the measured population distribution. Compare your simulation output with the results expected from the stochastic model.

2.10 Rather than the changes I made to the analytic model in equation (2.18) to "fix" the mismatch in figure 2.4, write simulation code that is a better representation of the processes encapsulated by equation (2.4). Compare the results of your code to analytic expectations.

2.11 What are the results of the following code?

```
int main(void)
{
    int i, j;
    i = 6; j = 2*(i++);
    printf("%d %d \n",i,j);
    i = 6; j = 2*(++i);
    printf("%d %d \n",i,j);
    return (0);
}
```

2.12 Which of the following are valid variable names: (a) `NotMe` (b) `I_am_OK` (c) `This-OK` (d) `howaboutthis?` (e) `ME` (f) `ain'tbad` (g) `_x37`

2.13 Remove a semicolon at the end of a line in the `bd.c` program, then try compiling it. What messages does the compiler give you?

2.14 Suppose that all residents can leave the system at any time (the processes of figure 2.1A instead of the processes of figure 2.1B examined throughout this chapter). Equation (2.6) instead becomes

$$\frac{dP_n}{dt} = \alpha P_{n-1} + (n+1)\beta P_{n+1} - (\alpha + \beta n)P_n.$$

Assuming immigration to extinct systems, show that at equilibrium

$$P_n = \frac{1}{n!}\left(\frac{\alpha}{\beta}\right)^n P_0.$$

Determine P_0, the mean, and the variance for this distribution. Modify `bd.c` to account for this new emigration process, measure the mean and variance and compare with your analytic results.

3
Logistic Birth–Death Models

Programming components covered in this chapter:
• global and local variables • functions • one-dimensional arrays • nested `for` loops • `scanf()` function

In this chapter we deal with birth and death processes within a closed ecological system, in contrast to the previous chapter's immigration and emigration model for an open ecological system. In a closed system, population changes occur only through births and deaths whose rates are functions of the number of individuals in the system.

After discussing its mathematical justification, I use the logistic growth function to describe births. On the one hand, the deterministic logistic growth model is wonderful in its mathematical brevity for conceptualizing the dynamics of a population that leads to a stable equilibrium. It enjoys widespread use in a variety of fields and provides a good fit to many types of empirical data (Banks 1994), primarily because it contains the most important terms of most functions. Also among its attractive features is that it can be solved analytically (see below). On the other hand, it lacks a solid conceptual foundation – even an accurate fit of data to a logistic growth model does not, in itself, yield any mechanistic understanding of the underlying processes.

The deterministic logistic growth model's conceptual vagueness arises from lumping birth and death processes into a single population growth term. In this chapter I compare deterministic and stochastic formulations of logistic growth and demonstrate differences between the two cases.

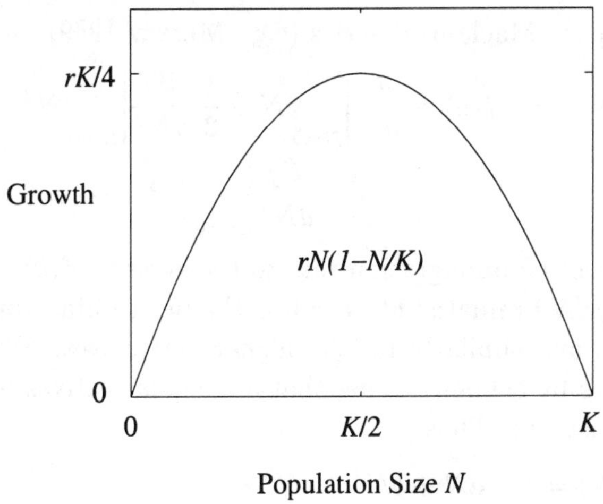

Figure 3.1. Form of the logistic growth function (equation (3.1)).

3.1 Origins of the Logistic Growth Function

While formulating their models of resource growth, many theoretical ecologists reach for a growth function of the logistic form

$$B(N) = rN\left(1 - \frac{N}{K}\right) = rN - \frac{r}{K}N^2, \tag{3.1}$$

where N is the population size, r is the per capita growth rate for very small populations and K is the carrying capacity of the closed system. The logistic function has two main features: At very low population densities the reproduction rate is proportional to the number of individuals, N, but as the population size increases, interspecific competition reduces the per capita reproduction rate. Its functional form is plotted in figure 3.1. The factor rN is responsible for the growth function going through zero at $N = 0$, but it also produces the roughly linear increase at low densities. At high densities growth is dominated by the factor $1 - N/K$, which is responsible for the growth function going through zero at $N = K$. In between these two zeroes is a maximum growth response at $N = K/2$.

Why is this functional form so popular? Ecologists don't actually believe that populations of organisms grow exactly according to this function. Instead, its popularity arises from its mathematical origins. Suppose that the *true* population growth function is represented by some completely unknown function $f(N)$. Most "well-behaved" functions[1] can be represented

[1] Well-behaved functions do not diverge to $\pm\infty$, and can be differentiated infinitely many times.

by a Taylor (or Maclaurin) series (e.g., Murray 1989)

$$\begin{aligned} f(N) &= f(0) + \left.\frac{df}{dN}\right|_{N=0} N + \frac{1}{2}\left.\frac{d^2f}{dN^2}\right|_{N=0} N^2 \\ &\quad + \frac{1}{2\cdot 3}\left.\frac{d^3f}{dN^3}\right|_{N=0} N^3 + \cdots , \end{aligned} \tag{3.2}$$

where $f(0)$ is the immigration rate to the system ($f(0) = 0$ if the system is closed), df/dN evaluated at $N = 0$ is the per capita growth rate of a small population, and similarly for the higher derivatives. We can clean up the Taylor series by acknowledging that all the derivatives evaluated at $N = 0$ are just constants. Thus

$$f(N) = I + aN + bN^2 + cN^3 + \cdots , \tag{3.3}$$

where the constants I, a, ..., represent the coefficients in equation (3.2). In this chapter I consider closed systems, hence $I = 0$, meaning there is no immigration or emigration. I also assume that competition exists between individuals for a limited resource, hence $b < 0$. This assumption means that interactions between a pair of individuals reduces population growth. Finally, I assume higher-order terms are unimportant, hence $c = 0$. What's left is a logistic function equivalent to equation (3.1).

3.2 Conceptual Vagueness of Logistic Growth

Arguments similar to those made above for growth could be made for the loss of organisms, leading to a Taylor series for an arbitrary mortality function. Assuming a closed ecological system again implies no emigration. If we include only the next two terms of the Taylor series expansion, loss is composed of a constant background mortality and a part that, most reasonably, increases with population size. These arguments imply a function $(c + dN)N$.

The following derivation illustrates the deterministic logistic model's conceptual vagueness. Using the outlined birth and death functions, I can write the population dynamics model as births minus deaths and collapse everything into a single logistic function

$$\begin{aligned} \frac{dN}{dt} &= \text{Births} - \text{Deaths} \qquad (3.4a) \\ &= (a_1N - b_1N^2) - (a_2N + b_2N^2) \\ &= (a_1 - a_2)N - (b_1 + b_2)N^2 \\ &= (a_1 - a_2)N - (a_1 - a_2)\frac{b_1 + b_2}{a_1 - a_2}N^2 \end{aligned}$$

$$= (a_1 - a_2)N\left(1 - \frac{b_1 + b_2}{a_1 - a_2}N\right)$$

$$= rN\left(1 - \frac{N}{K}\right). \tag{3.4b}$$

This procedure collapses four independent parameters, a_1, b_1, a_2, and b_2 into two aggregate parameters, $r = a_1 - a_2$ and $K = (a_1 - a_2)/(b_1 + b_2)$. The vagueness comes into play because its two parameters, r and K, both depend on more fundamental birth and death processes (e.g., Pielou 1969).

Many different processes lead to the same logistic growth model. As a specific example used throughout this chapter, consider the three sets of interactions (or models) in table 3.1 defined by the sets of four parameters listed in table 3.2.

Summing up the terms for any one of the models listed in table 3.1, for example model 2, gives the net rate of change of the population

$$\frac{dN}{dt} = \alpha N(1 - \frac{N}{2\kappa}) - \beta N(1 + \frac{\alpha N}{2\beta\kappa}) \tag{3.5}$$

$$= rN(1 - \frac{N}{K}), \tag{3.6}$$

where

$$r = \alpha - \beta \quad \text{and} \quad K = \frac{\alpha - \beta}{\alpha}\kappa.$$

Thus, all three models lead to the identical deterministic growth model (the same values for r and K), with no way of distinguishing between the various models, given experiments that measure only total rates of population change at specific densities.

However, details of individual interactions have important effects in the stochastic framework of individual-based simulations (Pielou 1969). These stochastic birth–death effects break the degeneracy (more than one model leading to the same result), or vagueness, of the deterministic logistic model. Such a detailed dependence could have important consequences for the interpretation of simulation results within broader theoretical constructs.[2]

3.3 Analysis of Deterministic Model

Equation (3.6) represents the general form for logistic growth, and is one of the few ecologically important differential equation models that can be

[2] Logistic models will be used in future chapters in situations that don't involve detailed stochastic treatments.

Table 3.1. Three distinct models that produce the same logistic model with $r = \alpha - \beta$ and $K = (\alpha - \beta)\kappa/\alpha$ (see equation (3.4b)).

Model	Birth interactions	Death interactions
1	$\alpha N(1 - \frac{N}{\kappa})$	βN
2	$\alpha N(1 - \frac{N}{2\kappa})$	$\beta N(1 + \frac{\alpha N}{2\beta\kappa})$
3	αN	$\beta N(1 + \frac{\alpha N}{\beta\kappa})$

Table 3.2. Connection between model parameters (α, β, and κ) and general birth–death model parameters (a_1, b_1, a_2, and b_2) for the models listed in table 3.1.

Model	a_1	a_2	b_1	b_2
1	α	β	$\frac{\alpha}{\kappa}$	0
2	α	β	$\frac{\alpha}{2\kappa}$	$\frac{\alpha}{2\kappa}$
3	α	β	0	$\frac{\alpha}{2\kappa}$

solved exactly. Accessibility to an analytic solution also makes it an ideal candidate for simulation models designed to test the effects of a multitude of ecological processes, such as spatial extensions, discreteness of individuals, environmental heterogeneity, and predation.

The most immediate result of equation (3.6) comes from an equilibrium analysis. Equilibrium analysis is an important tool in ecological theory, along with stability analysis, and is described in many introductory and advanced texts in ecology. Equilibrium implies a temporally constant state, and a verbal way of stating $dN/dt = 0$. This expression can only be satisfied for specific population densities N^* determined from the resulting condition

$$0 = rN^*(1 - \frac{N^*}{K}) \qquad \Longrightarrow \qquad N^* = 0, K. \tag{3.7}$$

Only one of these equilibria is really important for our analysis, $N^* = K$,

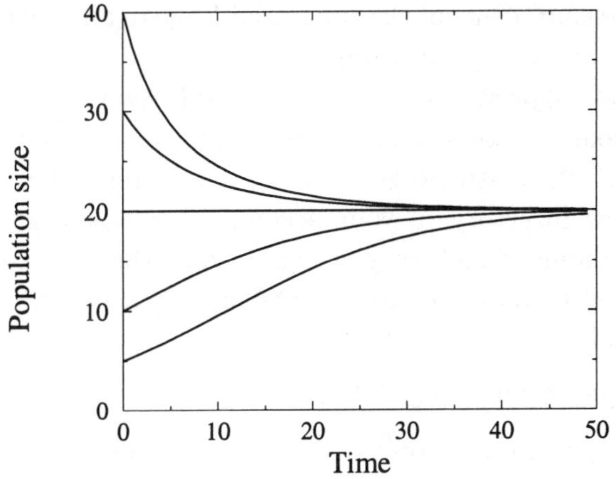

Figure 3.2. Dynamics of the logistic growth model for a variety of initial population sizes. In continuous-time, logistic growth is marked by the relaxation towards an equilibrium population size. ($\alpha = 0.2$, $\beta = 0.1$, and $\kappa = 40$, giving $r = 0.1$ and $K = 20$)

while the other simply states that an empty, closed system will not spontaneously generate individuals.

An exact solution to the logistic equation is obtained by multiplying both sides of equation (3.6) by dt then dividing by $N(1 - N/K)$, giving

$$rdt = \frac{dN}{N(1 - \frac{N}{K})}. \tag{3.8}$$

Solving equation (3.8) produces an explicit expression for the population as a function of time. Integrating the left side from 0 to t and the right side from $N(0) = N_0$ to $N(t)$ produces an expression for $N(t)$

$$N(t) = \frac{N_0 K}{N_0 + (K - N_0)e^{-rt}}. \tag{3.9}$$

Temporal dynamics of the logistic growth equation are shown in figure 3.2 for a variety of initial population sizes. A feature common to all runs is the relaxation to a stable equilibrium.

3.4 Logistic Birth–Death Simulation Model

Expanding on the C program in chapter 2, the following simulation tests the ability of the deterministic, logistic growth model to capture the dynamics of a stochastic, density-dependent birth–death process. As before, we will compare measurements of the simulation's mean and variance to a stochastic treatment's analytic predictions.

The following program introduces several new programming concepts, including global variables, arrays, and functions. It builds upon the previous stochastic birth–death program to a great extent. I recommend that you create a new directory in your account, perhaps **logbd**, for this project. From your home directory you could copy the program from chapter 2, **bd.c**, located in your directory **brthdth**, into a new file **logbd.c** using the cp command:

```
cp brthdth/bd.c logbd/logbd.c
```

Once the program is copied, you can start making edits to the new file rather than starting from scratch.

3.4.1 Code

```
     /* density dependent birth-death simulation */

     #include  <stdlib.h>

     #define   DELTAT          1.0
     #define   NPERT           ((int)(1.0/DELTAT))
     #define   MAXTIME    (1000*NPERT)
     #define   PRNTTIME   (10*NPERT)
     #define   PNTS       (MAXTIME/PRNTTIME)
     #define   ALPHA           0.2
     #define   KK         20.0
     #define   BETA       0.1

     /* Model 1 -- dN/dt = alpha N (1 - N / K) - beta N
        Model 2 -- dN/dt = alpha N (1 - N / 2K)
                     - beta N (1 + alpha N / 2K beta)
        Model 3 -- dN/dt = alpha N
                     - beta N (1 + alpha N / K beta) */

1.a) double    alpha, kk, beta;
2.a) int  PopChange(int, int);

     int main(void)
     {
1.b)      int  irun, ttt, n1, n2, n3, cnter, seed, numruns;
3.a)      double    pops1[PNTS], pops2[PNTS], pops3[PNTS];

1.c)      alpha = ALPHA*DELTAT;
          beta = BETA*DELTAT;
```

```
           kk = KK;
           seed = 1456739853;
           srand48(seed);

           printf("Enter number of runs:\n");
4)         scanf("%d",&numruns);

3.b)       for(ttt=0;ttt<PNTS;ttt++)
                pops1[ttt] = pops2[ttt] = pops3[ttt] = 0.0;

           for(irun=0;irun<numruns;irun++)
           {
                cnter = 0;
                n1 = n2 = n3 = 5;
                for(ttt=0;ttt<MAXTIME;ttt++)
                {
2.b)                 n1 += PopChange(n1,1);
                     n2 += PopChange(n2,2);
                     n3 += PopChange(n3,3);
                     if(ttt%PRNTTIME==0)
                     {
                          pops1[cnter]   += n1;
                          pops2[cnter]   += n2;
5)                        pops3[cnter++] += n3;
                     }
6)                   if(n1==0 && n2==0 && n3==0) break;
                }
           }
           for(ttt=0;ttt<PNTS;ttt++)
                printf("%d %f %f %f\n",ttt*PRNTTIME,
7)                        pops1[ttt]/(double)numruns,
                          pops2[ttt]/(double)numruns,
                          pops3[ttt]/(double)numruns);
           return (0);
      }

2.c) int PopChange(int pop, int modnum)
      {
1.e)       int temp=0, i;
           double xpop;

           xpop = (double)pop;
           if(modnum==1)
           {
                for(i=0;i<pop;i++)
1.d)                 if(drand48()<alpha*(1.0-xpop/kk)) temp++;
                for(i=0;i<pop;i++)
                     if(drand48()<beta) temp--;
           }
           if(modnum==2)
```

```
          {
               for(i=0;i<pop;i++)
                    if(drand48()<alpha*(1.0-xpop/kk/2.0)) temp++;
               for(i=0;i<pop;i++)
                    if(drand48()<beta
                         *(1.0+alpha*xpop/kk/beta/2.0)) temp--;
          }
          if(modnum==3)
          {
               for(i=0;i<pop;i++) if(drand48()<alpha) temp++;
               for(i=0;i<pop;i++)
                    if(drand48()<beta
                         *(1.0+alpha*xpop/kk/beta)) temp--;
          }
2.d)      return(temp);
     }
```

3.4.2 Code Details

Line 1: A few *global* variables are defined on line 1.a. In contrast, the type of variables we used last time and those on line 1.b, for example, are *local* variables. Local variables are defined within a routine, be it the `main()` routine or other function, or more generally within a compound statement. Local variables can only be accessed within the compound statement where they are defined. Global variables, such as **alpha**, can be referenced by those routines below their definition within the program file that contains the definition (chapter 9 deals with multi-file programs). Hence, the local variable **irun** is only defined for use within **main**, whereas the global variable **alpha** can be used throughout the program, for example, at lines 1.c and 1.d. Some programming books state that the use of global variables is sloppy, instead encouraging that all communication between routines be performed in more complicated ways. I use lots of global variables because they make communication between routines easier and my emphasis is on getting a scientific result.[3] Note the simultaneous declaration and initialization of **temp** on the same line (1.e).

Line 2: A *user-defined function* **PopChange()** is defined on line 2.a that implements stochastic versions of the three models defined in table 3.1. The function is defined in a function prototype statement as type **int**,

[3] Of course if you work on a commercial software product with hundreds of other programmers, then your global variable definitions might conflict with others' local variable definitions – that would be hazardous.

meaning it is treated like an integer variable and can be used on the right-hand side of an assignment statement, or anywhere else that an integer can be used. The one exception is that it can't be placed on the left-hand side of an assignment statement. In the definition statement on line 2.a, a function is denoted by the `()` at the end of the function name to distinguish it from the name of a regular variable. Also defined on this line are the data types of the arguments (or variables) passed to the function, in this case two arguments each of type `int`. Line 2.b demonstrates using `PopChange()`, explicitly showing two arguments – an `int` variable `n1` and the number 1 within the parentheses. Line 2.c marks the definition of the function as type `int` and its arguments, here called `pop` and `modnum`, of type `int`. Notice that the argument names at lines 2.b and 2.c are different. This difference is OK because the two names refer to different locations in the computer's memory, for example, one referring to the local variable called `n1` within `main` and the other local variable called `pop` within the function `pop1`. When arguments are passed, values of the calling function's variables are copied to the called function's variables. There are two important reasons to call them different names. First, it helps make the program clearer and easier to read. Second, interesting and subtle problems could occur when the name used in a local definition is the same as an accessible global variable (see exercise 3.14) and you and the compiler disagree on which of the two identically named variables your program references (the compiler always wins). Line 2.d demonstrates passing the returned value back to the calling routine. After the code in the function runs, the result held in the variable `temp`, dependent on the value of the argument `pop`, is sent back by the `return` statement. The variable in the `return` statement must be of the same type as the function itself.

PC Users: If your compiler fails to recognize `drand48()`, now might be a good time to turn the replacement mentioned in the footnote on page 30 into a user-defined function. I suggest the function `drand48pc()`

```
double drand48pc(void)
{
     return((double)rand()/(double)(RAND_MAX+1));
}
```

Line 3: Line 3.a defines three one-dimensional arrays, an example of which is `pops1`, short for "populations." An array definition consists of a name (just like the variable names used before), an open bracket, a number specifying the array size (or how many *elements* are desired), and a closing bracket. Each element is exactly like an individual variable used previously, depending, of course, on its type. In this case I use a series of precompiler expressions to specify the array size because the program needs this number in several places. If some time later a different number of elements is desired, all that is needed is one change at the precompiler definition. Arrays are very handy when there are many similar pieces of data. For example, `pops1` holds the population counts of model number 1 at various times. The program averages these numbers over many simulation runs, and when all the runs are finished prints out these numbers. It would be very inconvenient (but not impossible) to define a variable for each time, i.e., `pop1_0`, `pop1_1`, `pop1_2`, etc., as discussed in section 2.6: Arrays enable the mathematical equivalent of indexing a variable, x_i, where $i = 1, 2, ..., N$, allowing a much simpler way of writing code. The major caveat about arrays in C is that if you define an `array` with, say, 10 elements, the array elements run from `array[0]` to `array[9]`. You could certainly ask, "Why don't the element numbers run from 1 to 10?" In fact, FORTRAN and Pascal define elements beginning with 1, not 0. The reason C uses the indexing system it does is that it better reflects the addressing of memory by the computer, whereas starting with 1 reflects a more mathematical way of thinking. Line 3.b gives an example of how simple life becomes with arrays, using just a few lines of code to initialize all 100 elements of each array to zero.

Line 4: The `scanf()` function is one way for the user to communicate with the program while the program is running. It enables the input of new values for variables and subsequent actions by the program given those values. Arguments for `scanf()` are similar to the `printf()` statement. An important difference concerns fields used for `double` variables: use `%f` for `printf()`, but use `%lf` for `scanf()`. The reason is that `printf()` prints out `float` and `double` variables in the same way, but `scanf()` cannot treat the two identically. Also, `scanf()` is a function called by the program with the express purpose of changing the values of local variables, hence the presence of the ampersand (`&`) allows the `scanf()` function to access the variable in this intimate way.[4] Don't forget this ampersand!

Line 5: This line demonstrates updating an array element and the array's index. Expanding this statement out to its fullest it becomes

```
pops3[cnter] = pops3[cnter] + n3;
cnter = cnter + 1;
```

[4] `&` is the address operator giving the memory location of the variable (see section 4.1).

As mentioned before, the `+=` represents placing the variable on the left-hand side on to the right-hand side as well, and performing the indicated operation with the quantity located on the right-hand side. There are also operations, `*=`, `\=`, and `-=` among others. The incrementing performed on the index variable `cnter` occurs *after* being used in the expression, as indicated by the placement in the above expanded version of line 4.

Line 6: If in an `if` statement you specify two conditions that must hold true, the two conditions are separated by `&&`, which means "and." Also, `||` means "or," and `!` means "not."

Line 7: In the previous chapter we discussed the two basic types of variables, `int` and `double`, that you can use in your programs. It becomes a bit of a technical detail if you want to assign the value of a variable of type `double` to a variable of type `int`. For example, if `i` is an `int` variable and `x` is a `double` variable, then the results of the statement `i = x;` is unclear when `x = 1.3`. This conversion from one type of variable into another is usually performed implicitly by the compiler. However, the need often arises to coerce the compiler into *casting* variables from one type into another. Line 5 demonstrates a specific use of a casting operator to turn an `int` into a `double`. A common mistake is turning one `int` variable divided by another `int` variable into a `double` result, for example, if `x` is a `double` and `i = 10` and `j = 3` are both `int` variables. The statement `x = i/j;` yields a value `x = 3.0` rather than the perhaps more desirable value `x = 3.333333`.... The reason is that there is an implicit difference between `int` division and `double` division – the compiler assumes that the division of integers is going to be placed into an `int` variable and rounds the result down to the nearest integer. Thus, one needs to cast the `int` variables into `double` variables before the division takes place, or `x = ((double)i)/((double)j);`. (The extra sets of parentheses don't affect the result and help clarify the order of the operations (and they're free!)).

3.4.3 Preliminary Simulation Results

Data from the simulation are plotted in figure 3.3 for all three models at two different values for κ, $\kappa = 20$ and $\kappa = 40$, and averaged over three different numbers of runs, $M = 1, 10, 100$. In the case of single runs, figure 3.3A,B, there are no detectable differences between the three models (at least to my eyes!). Averaging over just ten runs in figure 3.3C,D indicate both a qualitative difference between the two carrying capacities and some

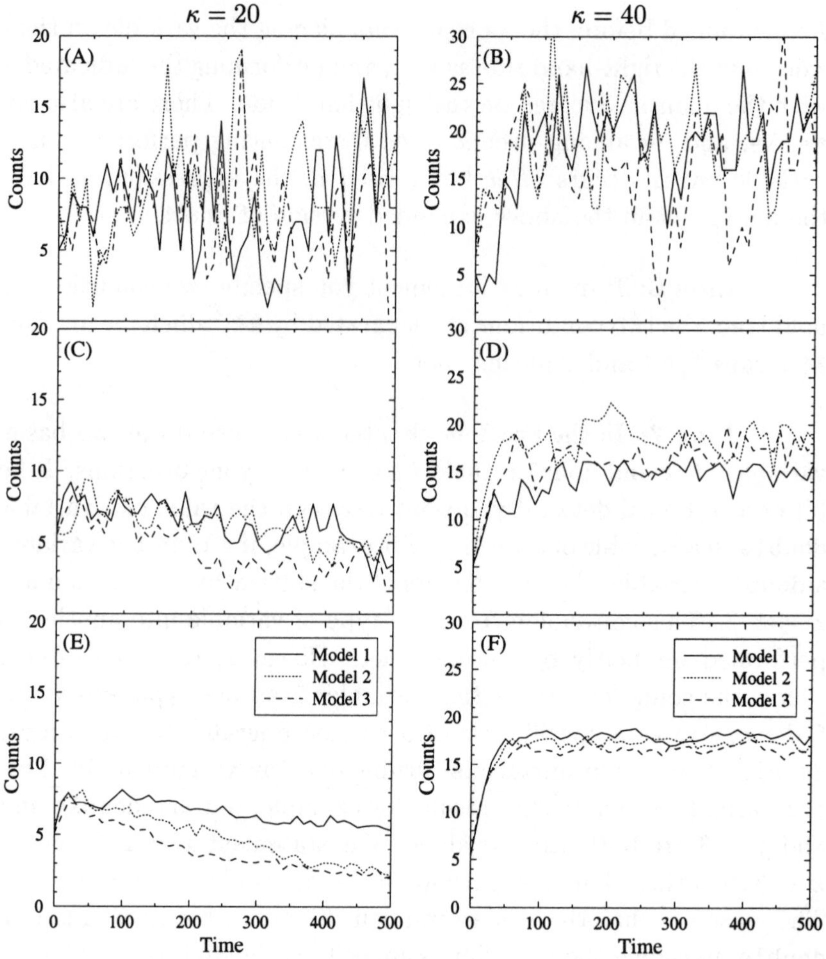

Figure 3.3. Model comparisons using $M = 1$ run (A,B), $M = 10$ runs (C,D), and $M = 100$ runs (E,F) and two κ values, $\kappa = 20$ (first column) and $\kappa = 40$ (second column). All panels use $\alpha = 0.2$ and $\beta = 0.1$. Counts for deterministically low population numbers decline when stochasticity is incorporated.

possible difference between the three models. Both of these features are greatly enhanced by averaging over $M = 100$ runs in figure 3.3E,F.

Why these differences? First, the downward slope seen in figure 3.3E occurs because many of the 100 replicates go extinct during the 1000 simulation steps. Since extinct replicates contribute zero to the total, the mean decreases. With a higher carrying capacity, figure 3.3F, extinctions are much

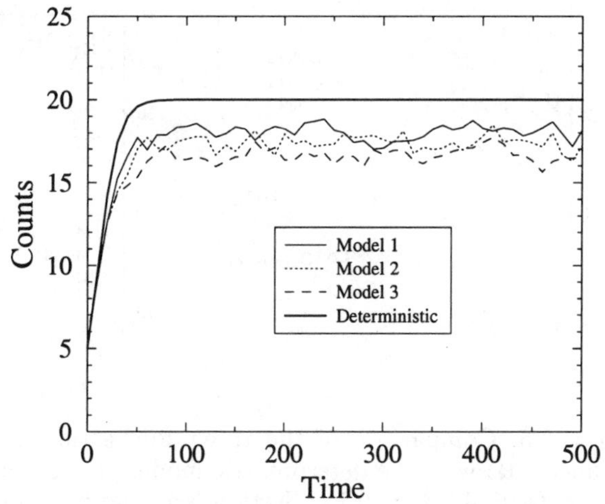

Figure 3.4. Comparison of results from the three stochastic models (figure 3.3F) with the deterministic model predictions for deterministically large population numbers ($\kappa = 40$, $M = 100$).

less likely, thus no decrease is detectable.[5] Second, figure 3.3E,F suggest that the long-term averages of the three models are different – the line for model 1 seems consistently above the line for model 2 which seems consistently above the line for model 3.

Equation (3.9) provides an analytic test of figure 3.3's simulation results – figure 3.4 plots the appropriate analytic results on top of the simulation results from figure 3.3F. Many features are qualitatively similar between the analytic and simulation results (in a rather generous light) such as the short transient rise in population density and stable equilibrium. However, the analytically predicted equilibrium density is clearly much too high.

We can make the comparison in a much more exhaustive manner by letting the computer take the averages over time rather than a simple visual inspection of the long-term behavior in figure 3.3. Figure 3.5A plots, as a function of the parameter α, the differentials between the theoretically predicted deterministic mean and simulation averages for the three simulation models. Each data point is an average over 200 runs. Each run has an initial population size set to the analytically predicted average and the initial 100 simulation steps are discarded. Measurements are then made once every 20 simulation steps for 2000 steps. Only nonzero populations are averaged.

[5] A detailed theoretical treatment shows that after an infinite time all of these systems will go extinct (see Nisbet and Gurney 1982).

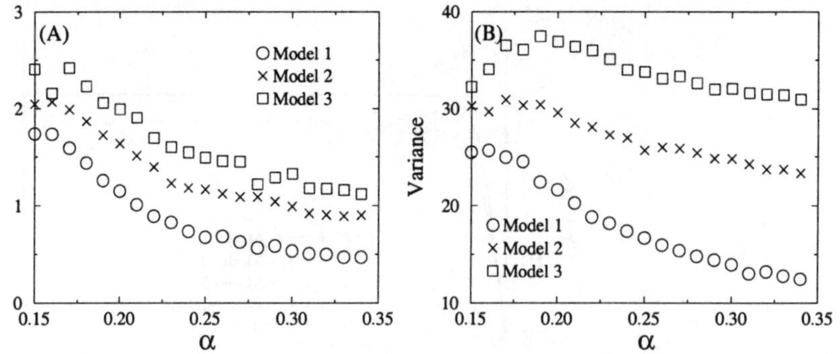

Figure 3.5. Comparison of the three stochastic models' means (A) and variances (B) with the deterministic model predictions for a range of birth rates α ($\kappa = 40$, $\beta = 0.1$). Section 3.5 examines the differences between the three stochastic models.

There is a small but consistent difference between the three simulation models, as well as with the deterministic model, for all values of α (figure 3.5A). Figure 3.5B plots the variances measured from the simulation data. There is, of course, no analytic estimate for the variance from a deterministic model, but the important feature is that differences exist between the model results.

3.5 Analytic Stochastic Birth–Death Model

The modeling approach covered in this section steps back from the differential equation model and considers the problem from a stochastic processes perspective. For the logistic model dealt with in this chapter, an analytic solution can be obtained even in the stochastic situation. Excellent presentations are given in Pielou (1969), Nisbet and Gurney (1982), and Renshaw (1991). Roughgarden (1975) performed a detailed analysis of the stochastic logistic model, and DeAngelis (1976) applied it to Canada goose population dynamics. Although for problems more general than logistic growth (more complicated birth and death functions) there seems to be little hope for exact solutions, the solution for the logistic example provides important insights.

Before analytically examining the stochastic model, we can get a feel for why we might anticipate differences between the three models. Figure 3.6 depicts the probability distributions, P_n, near $n = K$ for models 1 and 3. The arrows show the allowed population size transitions, and list the per capita rates at which transitions occur. In model 1, the birth rate decreases

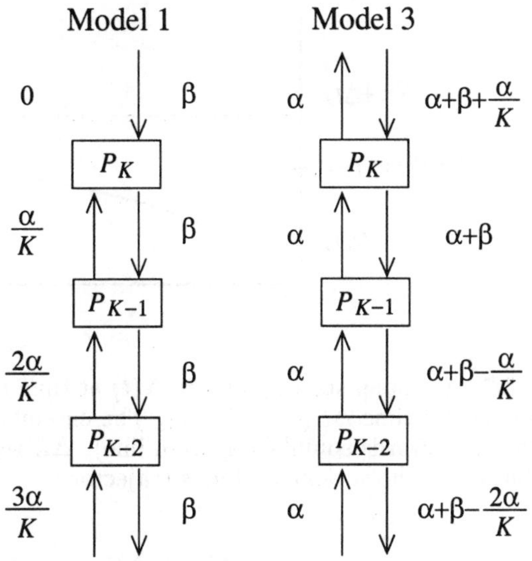

Figure 3.6. Differences between stochastic birth–death model formulations of models 1 and 3 in table 3.1 on page 54. Listed interactions (depicted as arrows) denote the per capita rates for transitions involving the gain or loss of one individual. In model 1, birth rates decrease to zero at the carrying capacity, whereas in model 3 the birth rate is independent of N but the death rate increases.

to zero at the carrying capacity K, meaning the population size can never exceed K (as long as it starts below K). In contrast, model 3 has a constant per capita birth rate, independent of population size, allowing the population to exceed K. The price is paid in an increased death rate as the population size increases. It is clear that there are fundamental differences between how the stochastic models work, thus the differences in the means and variances we observed should not be too surprising. In the next sections we pursue analytic models of stochastic processes that account for the differences.

3.5.1 Mean Population Size

Imagine that the system experiencing births and deaths has reached equilibrium[6] and we are examining the stochastic fluctuations about this equilibrium. Over a small, finite time interval Δt, during which multiple birth events (described by a function $B(N)$) and death events (described by a function $D(N)$) occur, we anticipate a population change from $N(t)$ (now

[6] Technically, the system reaches a *quasi-stationary equilibrium*, roughly defined as the equilibrium distribution of states *excluding* the extinct state (Nisbet and Gurney 1982).

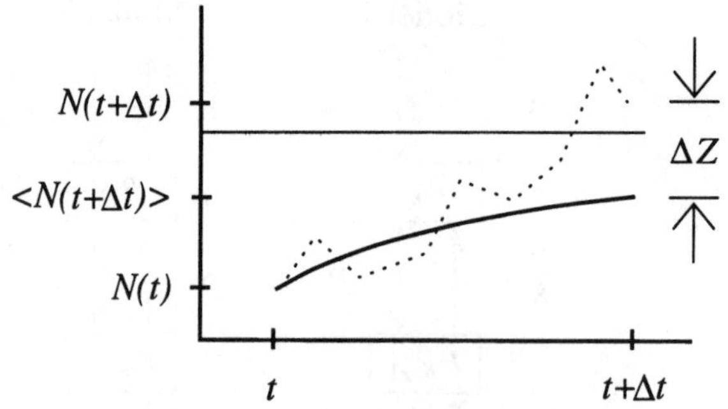

Figure 3.7. A system starting at size $N(t)$ at time t experiences stochastic events during a time Δt (dotted line). The ensemble average (heavy line), $\langle N \rangle$, relaxes towards equilibrium (thin line). ΔZ represents the stochastic contribution to the specific system's trajectory.

treated as a continuous variable) to $N(t + \Delta t)$ given by the expression

$$N(t + \Delta t) = N(t) + [B(N) - D(N)]\Delta t + \Delta Z. \tag{3.10}$$

Equation (3.10) is conceptually related to a discrete-time version of the birth–death equation (3.4a). The middle terms, $[B(N)-D(N)]\Delta t$, represent the expected deterministic change, and the last term, ΔZ, is the stochastic contribution to the population change over the time interval Δt. Figure 3.7 illustrates a good way to think of these various terms. Imagine running the simulation a million times, with each run initialized to $N(t)$, which is just below the carrying capacity (the horizontal line). On average, the runs will tend toward the carrying capacity, ending up at a value symbolized by $\langle N(t + \Delta t)\rangle$, representing the mean, average, or expectation value over the million runs. However, each run has its own trajectory, one of which is shown as the dotted line. The difference between a run's trajectory and the expected value is denoted as ΔZ.

More formally, consider an ensemble average of equation (3.10), which means averaging the expression over an ensemble of many independent replicates. The average, more formally called the expectation value, is represented by the brackets, $\langle \ldots \rangle$. Because additive terms are statistically independent, the ensemble average yields

$$\langle N(t + \Delta t)\rangle = \langle N(t)\rangle + (\langle B(N)\rangle - \langle D(N)\rangle)\Delta t + \langle \Delta Z\rangle. \tag{3.11}$$

Quickly this expression simplifies because of our assumption of the system being at steady-state, which implies $\langle N(t+\Delta t)\rangle = \langle N(t)\rangle$, cancelling on both sides. Likewise, $\langle \Delta Z\rangle = 0$ simply because stochastic fluctuations are defined to have an average of zero – otherwise the fluctuations would affect the mean and such a contribution should be accounted for in some deterministic manner. These cancellations leave

$$\langle B(N)\rangle = \langle D(N)\rangle. \tag{3.12}$$

I will now be a bit specific for $B(N)$ and $D(N)$ by using the general second-order functions of equation (3.4)

$$\begin{aligned} B(N) &= a_1 N - b_1 N^2 && (3.13a) \\ D(N) &= a_2 N + b_2 N^2, && (3.13b) \end{aligned}$$

that embody all the models listed in table 3.1 through the conversions listed in table 3.2. Equation (3.12) then produces

$$\begin{aligned} \langle (a_1 N - b_1 N^2)\rangle &= \langle (a_2 N + b_2 N^2)\rangle, && (3.14a) \\ a_1\langle N\rangle - b_1\langle N^2\rangle &= a_2\langle N\rangle + b_2\langle N^2\rangle, && (3.14b) \\ (a_1 - a_2)\langle N\rangle &= (b_1 + b_2)\langle N^2\rangle. && (3.14c) \end{aligned}$$

If we now define the mean, $m = \langle N\rangle$, and variance, $V(N) = \langle N^2\rangle - m^2$, of the population then equation (3.14c) gives

$$\begin{aligned} (a_1 - a_2)m &= (b_1 + b_2)(V(N) + m^2) \\ m^2 &= \frac{(a_1 - a_2)m}{b_1 + b_2} - V(N) \\ m &= \frac{a_1 - a_2}{b_1 + b_2} - \frac{V(N)}{m} \\ &= \frac{r}{s} - \frac{V(N)}{m} \\ &\approx \frac{r}{s} - \frac{sV(N)}{r}, && (3.15) \end{aligned}$$

where $r = a_1 - a_2$, $s = b_1 + b_2$. The approximation assumes a small variance-to-mean ratio in comparison to the deterministic mean r/s. In the sense of equation (3.15), the variance resulting from a stochastic process reduces the expected mean from its deterministic value. As seems quite reasonable for the deterministic limit, $V(N) = 0$, the stochastic mean collapses to the deterministic expectation.

3.5.2 Variance of Population Fluctuations

Equation (3.15) shows that the population mean depends on the variance about the mean, and we will now calculate the variance following Pielou's

(1969) concise derivation. If we represent the population number at some time t as

$$N(t) = [1 + u(t)]m, \tag{3.16}$$

then $u(t)$ is a small fractional displacement from the stochastic mean m with the important property $\langle u \rangle = 0$. Also note that the variance in N and u are related by $V(N) = m^2 V(u)$. We begin by rewriting the basic equation for a stochastic process for our specific problem

$$N(t + \Delta t) = N(t) + \left(rN - sN^2\right)\Delta t + \Delta Z, \tag{3.17}$$

where $r = a_1 - a_2$ and $s = b_1 + b_2$ as above. Substituting equation (3.16)

$$\begin{aligned} [1 + u(t + \Delta t)]\, m &= [1 + u(t)]\, m \\ &\quad + \left\{rm\,[1 + u(t)] - sm^2\,[1 + u(t)]^2\right\}\Delta t + \Delta Z. \end{aligned} \tag{3.18}$$

I now introduce a shorthand notation to make things slightly cleaner, $u \equiv u(t)$ and $\Delta u \equiv u(t + \Delta t) - u(t)$. Then

$$\begin{aligned} (1 + u + \Delta u)m &= (1 + u)m + \big[rm(1 + u) \\ &\quad - sm^2(1 + u)^2\big]\Delta t + \Delta Z \\ (1 + u + \Delta u) &= (1 + u) + \big[r(1 + u) \\ &\quad - sm(1 + 2u + u^2)\big]\Delta t + \frac{1}{m}\Delta Z \\ u + \Delta u &= u + \big[r(1 + u) - sm(1 + 2u + u^2)\big]\Delta t + \frac{1}{m}\Delta Z \\ &= u + (r - ms)\Delta t + (r - 2sm)u\Delta t \\ &\quad - smu^2\Delta t + \frac{1}{m}\Delta Z. \end{aligned} \tag{3.19}$$

Now approximate the stochastic mean m by the deterministic mean, $m \approx r/s$, and assume that the fractional displacement u is small ($u \ll 1$) such that $u^2 \approx 0$, giving

$$u + \Delta u = (1 - r\Delta t)u + \frac{s}{r}\Delta Z. \tag{3.20}$$

Squaring both sides

$$(u + \Delta u)^2 = \left((1 - r\Delta t)u + \frac{s}{r}\Delta Z\right)^2 \tag{3.21}$$

$$= (1 - r\Delta t)^2 u^2 + 2(1 - r\Delta t)\frac{s}{r}u\Delta Z + \left(\frac{s}{r}\right)^2 \Delta Z^2, \tag{3.22}$$

then taking expectations over an ensemble of systems

$$\langle (u+\Delta u)^2 \rangle = (1-r\Delta t)^2 \langle u^2 \rangle + 2(1-r\Delta t)\frac{s}{r}\langle u\Delta Z\rangle + \left(\frac{s}{r}\right)^2 \langle \Delta Z^2 \rangle, \tag{3.23}$$

gives us a messy equation that simplifies after thinking about it for awhile. First, the equilibrium assumption means that the variance is constant in time, $V(u) = \langle (u+\Delta u)^2\rangle = \langle u^2 \rangle$. If we also assume Δt is small, then, $(1-r\Delta t)^2 \approx 1 - 2r\Delta t$, giving

$$0 = -2r\Delta t\langle u^2\rangle + 2(1-r\Delta t)\frac{s}{r}\langle u\Delta Z\rangle + \left(\frac{s}{r}\right)^2 \langle \Delta Z^2\rangle. \tag{3.24}$$

Next, we assume that stochastic fluctuations are independent of the system's state, or $\langle u\Delta Z\rangle = \langle u\rangle\langle \Delta Z\rangle$. Since both $\langle u\rangle$ and $\langle \Delta Z\rangle$ are zero, the second term is zero, giving

$$0 = -2r\Delta t\langle u^2\rangle + \left(\frac{s}{r}\right)^2 \langle \Delta Z^2\rangle$$

$$2r\Delta t\, V(u) = \frac{s^2}{r^2} V(\Delta Z) \tag{3.25a}$$

$$V(N) = \frac{1}{2r\Delta t} V(\Delta Z). \tag{3.25b}$$

The question (via equation (3.25b)) becomes what is $V(\Delta Z)$? We derive that now.

Consider our system at equilibrium, or, in other words, when it has N individuals where N is defined by $B(N) = D(N)$. During a short time interval Δt two things might happen – the system gains a new individual with probability $B(N)\Delta t$ giving $\Delta Z = +1$, or the system might lose an individual with probability $D(N)\Delta t$ giving $\Delta Z = -1$. Of course, another possibility is that nothing happens, meaning the system retains exactly N individuals with probability $1 - [B(N) + D(N)]\Delta t$ giving $\Delta Z = 0$. Making use of the general definition for the expectation value

$$\langle \text{value} \rangle = \sum_{\text{possible events}} (\text{event value}) \times (\text{event probability}), \tag{3.26}$$

our specific event probabilities imply that the mean of the stochastic portion of equation (3.10) is

$$\begin{aligned} \langle \Delta Z \rangle &= (+1)\times B(N)\Delta t \;+\; (-1)\times D(N)\Delta t \\ &\quad + (0)\times[1-(B(N)+D(N))\Delta t] \\ &= (B(N)-D(N))\Delta t = 0, \end{aligned} \tag{3.27}$$

and similarly for the variance of the stochastic portion

$$\begin{aligned}\langle \Delta Z^2 \rangle &= (+1)^2 \times B(N)\Delta t \;+\; (-1)^2 \times D(N)\Delta t \\ &\qquad + \; (0)^2 \times [1 - (B(N) + D(N))\Delta t] \\ &= [B(N) + D(N)]\Delta t \;= \mathrm{Var}(\Delta Z). \end{aligned} \tag{3.28}$$

Putting together equations (3.13), (3.16), (3.25b), and (3.28) gives the variance in the population at equilibrium

$$\begin{aligned} V(N) &= \frac{1}{2r\Delta t}[B(N) + D(N)]\Delta t \\ &= \frac{1}{2r}[(a_1 N - b_1 N^2) + (a_2 N + b_2 N^2)] \\ &= \frac{1}{2r}[(a_1 + a_2)N + (b_2 - b_1)N^2] \\ &\approx \frac{1}{2r}\left[(a_1 + a_2)(r/s) + (b_2 - b_1)(r/s)^2\right] \\ &= \frac{1}{2}\frac{a_1 + a_2}{b_1 + b_2}\left(1 + \frac{(a_1 - a_2)(b_2 - b_1)}{(a_1 + a_2)(b_1 + b_2)}\right). \end{aligned} \tag{3.29}$$

Because the three models have different values for b_1 and b_2 (see table 3.2 on page 54) we can expect different variance measurements from the simulations. These different variance measures also feed back through equation (3.15) into different means for the three simulation models.

3.5.3 Comparison of Simulation and Analytic Models

Figure 3.8 compares simulation results with the above stochastic model predictions. Simulation data are from figure 3.5 and the line closest to each data set is the analytic prediction. Predictions from the stochastic model are in very good agreement with the simulation results; however, significant discrepancies exist – particularly for model 3. I leave it as an exercise to explore potential sources for this mismatch using your simulation.

Stochastic models for all kinds of ecological interactions have been examined (Bartlett 1960, Goel and Richter-Dyn 1974), including competition (Leslie and Gower 1958), predation (Leslie and Gower 1960), competition with predation (Poole 1974), and disease epidemics (van Herwaarden 1997). The interesting, early works represent the foundation in ecology of connecting individually explicit simulation models with analytic formulations.

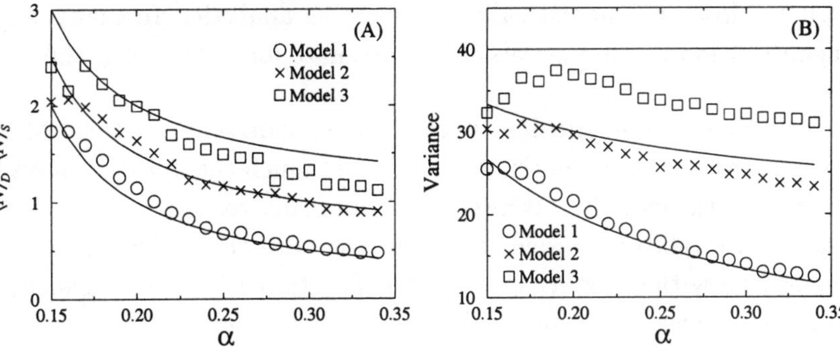

Figure 3.8. Comparison of the three simulation models' means and variances (figure 3.5) with the stochastic model predictions ($\kappa = 40$, $\beta = 0.1$). Agreement is quite good excepting the case of model 3's variance (see exercise 3.13).

3.6 Review of Modeling Concepts

In the last two chapters I showed examples of several different types of modeling frameworks and concepts related to them, all of which you are likely to see throughout the literature of theoretical ecology and evolution. Here I briefly review these ideas and their relationships as explored up to this point in the book.

Simulation models versus analytic models. Simulation models are the most straightforward representation of stochastic processes. As the programs of the last two chapters demonstrated, we need no appeal to mathematics to put them into the computer and run them; all we need is a set of rules describing or defining the interactions between individuals. The hard part comes when trying to analyze and understand the simulation results with the goal of advancing the scientific understanding of ecological systems. Simulation results with no appeal to broader theoretical results are like experimental measurements with uncalibrated instruments. Without calibration the experiments are meaningless; likewise without appeal to a theoretical foundation, simulations have too many ingredients that could be responsible for observed phenomena. Indeed, complicated simulations are much like an ecological simulation – a broader theoretical synthesis needs to be done to analyze their results.

Analytic models serve the purpose of grounding simulation models. I am rather generous in my definition of an analytic model; anything that can be written down mathematically I classify as analytic. In many ways what I mean is a model that serves as an alternate formulation to the simulation.

Stochastic models versus deterministic models. Most biological processes are inherently stochastic, yet much theoretical analysis involves deterministic models. Why? An important reason is that deterministic models are relatively easier to solve (compare the logistic growth model of section 3.3 with the stochastic model of section 3.5). Beyond such pragmatic reasons, a reasonable, inherent, and often justifiable assumption contained in a deterministic model is that a biological system consists of a large collection of individuals experiencing the same ecological interactions. This assumption implies that for any particular measure of the system's state, the dynamics of the measure's average, or mean, is a sufficient description. In other words, the variance has a negligible influence.

Many so-called mean-field models express intuited dynamical processes, the logistic equation (3.4b) being one example. However, we saw that analyzing a more detailed model accounting for stochasticity leads to different quantitative predictions (equation (3.15)). Recognizing that ecological interactions always have stochastic elements, do we always need to consider stochastic effects? This is a hard question to answer. Are we hoping for quantitative or qualitative predictions to experimental systems? Are there interesting phenomena arising from stochastic processes, or does stochasticity just make the mathematics harder with no new biological insight? We must trade off mathematical precision (and concommitant mathematical difficulties) and the biological essence we want reflected in a theoretical model.

A realization versus an ensemble average. These terms embody very important ideas for the connection between simulation models and analytic models. A single run, for example one of the lines displayed in figure 3.3A,B, represents a realization of a stochastic model. Because the temporal progression depends on random numbers, there is no hope for an analytic model to exactly reproduce a single simulation run. However, averaging over many runs produces an ensemble average, reducing the stochastic effects and leaving repeatable ones like the measures shown in figure 3.3E,F. We can be more hopeful in constructing analytic models for the averages. Unfortunately, ecological data are often in the form of a single population time series, some lasting many decades, representing a single realization of an ecological system. The challenge is to extract mechanistic insight from a single realization of a stochastic process.

Discrete events versus continuous processes. Within an experimental population of organisms experiencing only a constant death rate, it might be that after the start of the experiment the first individual dies at time $t = 4.1$ hours and a second at $t = 6.3$ hours. In contrast, if we had 10^{100} experimental populations, then we can easily imagine that the average number of individuals per collection behaves like a negative exponential, changing by very small amounts over very small time intervals. What is the best way of describing this system? Are distinct individuals dying at specific points in time, or is the population experiencing a constant loss in density? Both are clearly correct, but the usage depends on whether we are discussing realizations or ensemble averages. In a simulation, interactions between individuals are discrete events that occur at specific times. However, any useful analytic model of a population measure will involve continuous processes.

3.7 Exercises

3.1 Input the program for the density-dependent birth–death model and examine the dynamics obtained by averaging over 1 run, 10 runs, and 100 runs as in figure 3.3. Test whether differential extinction rates during the initial transient are important for the $\kappa = 40$ case by measuring the number of extinct replicates.

3.2 Obtain equation (3.9) from equation (3.8).

3.3 Write a program that determines how many prime numbers there are between 1 and 1 000 000. How many are there (include 1 as a prime)?

3.4 Which two lines of code are the same?

```
a)   x = a + b * c + d / e;
b)   x = (a + b) * (c + d) / e;
c)   x = (a + (b * c) + (d / e));
d)   x = (a + (b * c) + d) / e;
```

3.5 You have a program with two `double` variables, `x` and `y`. You want to exchange their values. How do you do it without losing either value?

3.6 Write a program to calculate $N(t)$ for the discrete-time and continuous-time equations. Compare the results for $\Delta T = 10.0, 1.0, 0.1$ and compare these results with those of the simulation. Explain any differences.

3.7 This exercise compares three distinct logistic growth processes specified by their values for a_1, a_2, b_1, and b_2: (A) {0.1,0,0,0.1/20}; (B) {0.2,0.1/20,0.1,0}; and (C) {0.2,0,0.1,0.1/20}. (a) Using equations (3.15) and (3.29), write a program to calculate the expected values for the mean and variance of the three logistic models. (b) Modify the birth–death simulation code to calculate the mean and variance of the population numbers over an arbitrary time interval using the algorithm described in exercise 2.4. Compare your simulation output with the expected results from part (a).

3.8 Add a cubic term onto the logistic model. Determine the effect on the mean and variance. Program it.

3.9 Write a program that converts an arbitrary number of seconds into 365-day years, 30-day months, weeks, days, hours, minutes, and remaining seconds.

3.10 Is there anything wrong here?

```
one_var + two_var = three-var;
```

3.11 Using arrays, modify the `bd.c` program of chapter 2 to simulate a closed system of many patches with the individuals moving between patches (see Roff 1974a, 1974b). Given that all patches are accessible from all other patches, how does the distribution of individuals per patch change with the number of patches? (Note: Make sure to keep the average number of individuals per patch constant.)

3.12 What is the output from these lines of code?

```
int main(void)
{
        int i1=2, i2=4, i3=5, ii;
        double x1=2.0, x2=4.0, x3=5.0, xx;
        ii = i3/i1; xx = i3/i1;
        printf("%d %f\n", ii, xx);
        ii = x3/x1; xx = x3/x1;
        printf("%d %f\n", ii, xx);
        ii = i2/i3; xx = i2/i3;
        printf("%d %f\n", ii, xx);
        ii = x2/x3; xx = x2/x3;
        printf("%d %f\n", ii, xx);
        return (0);
}
```

3.13 Shrink Δt in the simulation and repeat the comparison of figure 3.8. Does a small Δt improve agreement with analytic predictions?

3.14 Local and global variables sometimes conflict. Enter the following code into a file, compile and execute it. What happens?

```
int i;
void MyFunc(void);
int main(void)
{
        i = 5;
        printf("main1: %d\n", i);
        MyFunc();
        printf("main2: %d\n", i);
        return (0);
}
void MyFunc(void)
{
        int i;
        i = 10;
        printf("MyFunc: %d\n", i);
}
```

3.15 Predict what is produced by the following code.

```
int main(void)
{
        int i = 5;
        printf("main1: %d\n", i);
        i = i++;
        printf("main2: %d\n", i);
        return (0);
}
```

4
Random Numbers and Visualization

Programming components covered in this chapter:

- pointers
- passing function arguments by address
- multidimensional arrays
- writing to external files
- PostScript files

Let me explain why these two disparate topics – random numbers and visualization – belong in the same chapter. Neither topic involves ecological theory per se, but both are extremely important for simulating ecological and evolutionary systems, particularly in the context of individual-based, spatially explicit models. Simulation results are critically dependent on the soundness of your random number generator – in this chapter I present simulation results obtained with a really poor random number generator and compare them to results using acceptable generators. Similarly, understanding simulation results of complex ecological systems can be enhanced with the proper visualization tools. Rather than present visualization within a framework of ecological theory, I combine these two topics and demonstrate making and using Encapsulated PostScript files to visualize the problems of the poor random number generator. We will use PostScript files extensively in later chapters.

I will also demonstrate the basics of using system-supplied random numbers to produce arbitrarily distributed random numbers. For example, a commonly encountered distribution of random events is the normal distribution, but system-supplied random numbers are usually only provided uniformly over some interval. Mathematical analysis can provide a one-to-one correspondence between these two distributions, giving an efficient means of simulating arbitrarily distributed stochastic processes.

4.1 Pointer Variables

Pointer variables constitute an important technical component of the C programming language, and I use them a great deal in the rest of the book's programs. They provide a degree of confusion to new and old programmers alike, but their utility far outweighs the frustration of understanding them. The two most important reasons to use pointers involve communication between calling and called functions, and array manipulation.

4.1.1 Variables and Memory Locations

When a variable is defined, for example

```
int ttt;
```

the declaration tells the compiler to set aside a location in memory for the variable (see figure 4.1). Every location of the computer's memory has an address (think of these addresses as hexadecimal numbers[1]).

Thus, the word in the memory that the compiler associates with the label `ttt` has an address[2] which the compiler uses instead of the label. In short, a computer uses an address for a variable, not a character name. An assignment statement tells the computer to push a specific number into the address corresponding to the referenced variable. C lets you find out the address of a variable through an *address operator*, the ampersand symbol `&`, which "operates" on a variable and returns its address.

C allows for the declaration of variables that hold addresses; these variables are called "pointers" because the addresses they contain *point* to memory locations that hold information.[3] Pointer names look just like variable names, but are declared with a `*` such as,

```
int *pnter;
```

Associate the `*` with the word `int`, as in `(int *)pnter` when thinking about the pointer. You can read this two ways, either "`pnter` is a variable of type `(int *)`", or "`pnter` is a pointer to type `int`". However you choose to read it, when the compiler encounters this specific statement, it allocates memory to hold the address of an integer variable. A word of caution –

---

1 An efficient way of expressing binary numbers is using hexadecimal numbers, counting in base 16 instead of base 10 as we usually do, by using the letters `A` through `F` for the numbers 10 through 15. Hence, the hexadecimal number `A5` is equal to the decimal number 165: $10 \times 16 + 5 \times 1 = 1 \times 100 + 6 \times 10 + 5 \times 1$.

2 The compiler actually defines an offset from the start of the compiled code.

3 You can also define pointers to pointers (see the example on page 202).

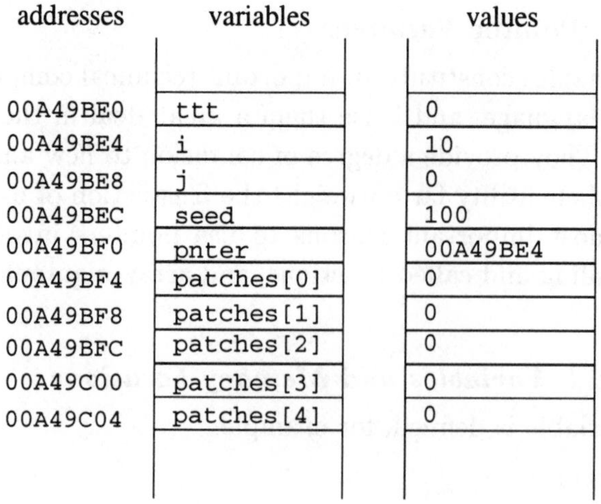
addresses
variables
values
00A49BE0
ttt
0
00A49BE4
i
10
00A49BE8
j
0
00A49BEC
seed
100
00A49BF0
pnter
00A49BE4
00A49BF4
patches[0]
0
00A49BF8
patches[1]
0
00A49BFC
patches[2]
0
00A49C00
patches[3]
0
00A49C04
patches[4]
0

Figure 4.1. Relationship between memory addresses, the variables associated with them, and the values those variables hold. C allows for variables, called pointers, that hold memory addresses. `pnter` is one such variable and has been assigned the address of `i`.

the pointer's value is undefined until the programmer assigns the address of some variable to the pointer. Uninitialized pointers usually point to sensitive areas of memory, and if the pointer is used, bad things like program crashes can occur. Finally, there can be lots of confusion about variables, pointers, and even pointers to pointers, and I often minimize confusion by naming pointers incorporating the suffix `pnter`.

Consider the following lines of code

```
int ttt, i, j, seed, *pnter, patches[NUM_ELEMS];
ttt = j = 0; i = 10; seed = 100;
pnter = &i;
```

in conjunction with figure 4.1. The first line of code sets aside space in memory for the declared variables in a one-to-one manner. Three important details are involved. First, an 8-bit[4] byte is the smallest unit of addressable memory even though the word size of your computer is likely 32 bits. Second, an integer variable takes up 4 bytes = 32 bits = 1 word. Finally, remember that a variable of type `double` stretches across two words, taking up the memory space of two integer variables. Also, recognize that the address of any variable is the address of its first byte.

[4] Remember, a bit is an entity taking on the values 0 or 1.

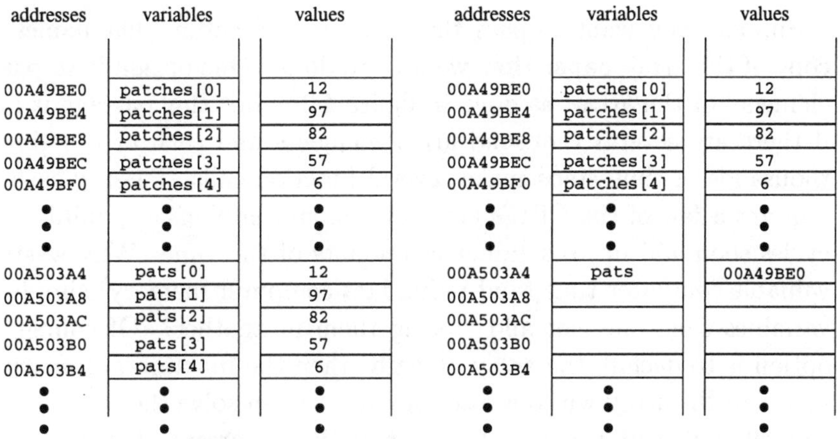
addresses
variables
values
00A49BE0
patches[0]
12
00A49BE4
patches[1]
97
00A49BE8
patches[2]
82
00A49BEC
patches[3]
57
00A49BF0
patches[4]
6
00A503A4
pats[0]
12
00A503A8
pats[1]
97
00A503AC
pats[2]
82
00A503B0
pats[3]
57
00A503B4
pats[4]
6
addresses
variables
values
00A49BE0
patches[0]
12
00A49BE4
patches[1]
97
00A49BE8
patches[2]
82
00A49BEC
patches[3]
57
00A49BF0
patches[4]
6
00A503A4
pats
00A49BE0
00A503A8
00A503AC
00A503B0
00A503B4

(A) Inefficient (B) Efficient

Figure 4.2. Alternative ways of passing arrays to functions. Memory associated with the calling function is at the top, and memory associated with the called function is shown at the bottom. (A) You can inefficiently pass data to the called function, requiring the CPU to copy every array value to the function's memory. (B) The efficient way to pass an arrays-worth of data is to pass only the array's base address to a pointer defined in the function.

The second line of code initializes some of the variables, throwing the specified values into the appropriate memory locations (which means setting the bits so as to represent the specific numbers).[5]

The third line of code is the relevant one to our discussion of pointers. On the right-hand side of the assignment statement, the address operator acts on the variable `i`, returning its address, `00A49BE4`. This address is placed into the memory location set aside for the pointer variable, `pnter`. Thus, a pointer is a completely different beast than an `int` or `double` variable. Likewise, `int` and `double` pointer variables are just as distinct as the variables themselves – do not put addresses of `double` variables into `int` pointers.

4.1.2 Using Pointers

Pointers are intimately tied to arrays. In the above example, the compiler allocates space for `NUM_ELEMS` array elements as required by the array `patches`. This allocation is assured to be a contiguous block of memory, as suggested in figure 4.1.

[5] Some compilers automatically initialize array elements to zero when the array is defined – either check out what your compiler does or be on the safe side and explicitly initialize your array elements.

Suppose you want to pass this array to a function that names its local copy of the array `pats`. How would you do it? One option is to pass all the elements of the array, as done in figure 4.2A, but this passage is inefficient if there are several thousand array elements and each of the simulation's thousands of time steps make several function calls. Passing each element requires a few of the CPU's clock cycles, and millions and millions of clock cycles soon add up to significant amounts of run time. Why waste all this valuable computer time (and redundant computer memory) simply copying variables from one slot and storing them in another? Of course, another option is to declare the array globally, then the function automatically has access to the array without any copying. I often solve the problem this way. But what if your program has several similar arrays thay you want your function to operate on at some time or another? How would you tell the function which of the arrays to work on during any specific function call? The answer is that the function only needs to know the array's starting location in memory (the address of the variable `patches[0]` in figure 4.2B) to access all of the array's elements. For example, consider the following short piece of code

```
     int main(void)
     {
          int patches[NUM_ELEMS];
          ...
          DoFunction(patches);
/*        DoFunction( &(patches[0]) );  Alternate Form */
          ...
     }

     void DoFunction(int pats[NUM_ELEMS])
/*   void DoFunction(int pats[])        Alternate Form */
/*   void DoFunction(int *pats)         Alternate Form */
     {
          ...
     }
```

Whichever format you choose to write this code in, the act of passing the array to the function is accomplished via passing the array's pointer – the address of the first element. This transfer is done very quickly since the only thing copied to `DoFunction`'s local variables is a single address. Note that the compiler doesn't care what value you place inside the brackets for the definition of `pats[NUM_ELEMS]`; you could have dropped the number of elements out entirely, i.e., `pats[]`. In either case the important aspect is that you declare the argument as a pointer, which, as a third option, you could do explicitly.

Once having passed an array's address to a function, you can get yourself into a bit of trouble. The compiler trusts you to know so completely what you are doing that nothing prevents you from accessing the array element `pats[2*NUM_ELEMS]` even though this element does not exist in the context of the array defined as `patches` in the `main()` routine. That memory address could be a part of your program, or someone else's program, and if you write to it expect a message like, `Segmentation fault -- core dumped.`[6] Make sure you pass valid addresses to your functions!

How do you deal with a pointer once you have one? Assume we defined a pointer variable `int *pats`. We can assign the pointer the address of the first element of an integer array, called the base address, with the statement

```
pats = &patches[0];
```

There are now a few useful things we can do using the pointer. First we can set the value of `array[0]` using the pointer

```
*pats = 12;
```

In this statement the * operator, technically called the indirection operator, acting on the pointer accesses the memory location pointed to by the pointer. When thinking about what the pointer points to, you should associate the * with the pointer name, as in `int (*pats)` and you see that the entity "`(*pats)`" is of type `int`.

As specific examples of how you can use pointers in conjunction with arrays, the following blocks of code are equivalent

```
for(i=0;i<10;i++) patches[i] = 0;
```

is the same as

```
pats = &patches[0];
for(i=0;i<10;i++) pats[i] = 0;
```

is the same as

```
pats = patches;
for(i=0;i<10;i++) *(pats + i) = 0;
```

is the same as

```
pats = patches;
for(i=0;i<10;i++) *(pats++) = 0;
```

[6] Addressing array elements beyond the array's defined boundaries is a notorious and common "bug." The computer will sometimes suffer silently but give you false results, and sometimes complain but not give any useful information as to the source of the problem. Keep this possibility in mind (and examine array index values) as a source of program crashes.

is the same as

```
pats = patches;
for(i=0;i<10;i++,pats++) *pats = 0;
```

The first block is self-explanatory. The second block first initializes the pointer to the base address of the array, then treats `pats` just like an array within the `for` loop. The deeper meaning is that any pointer can be treated just like the base address of an array, and we can add an index that the computer will use to calculate an offset. It is our responsibility as programmers to ensure that the pointers do not point to sensitive areas of memory. The third block of code initializes `pats` to `patches`, which simply drops off the entire indexing portion of the array. This demonstrates the equivalence between the concepts of pointers and arrays – dropping the index from an array yields a pointer and adding an index to a pointer yields an array. Within the `for` loop, the base and offset are explicitly added rather than having the computer adding them together implicitly through indexing as in the second block. The `for` loop of the fourth block reminds us that we can change the value of `pats` whenever we wish, and in this case we increment it after using it so that it points to the next array element after each usage. Finally, the last block of code increments the pointer with the loop variable, thereby cleaning up the assignment statement.

4.2 Basic Random Numbers

There is a saying, "Random numbers are far too important to be left to chance."[7] In simulation work this statement is very true: The construction of code producing random numbers is best left to professionals (but check out the resulting numbers just in case).

All random numbers created by computers are not random. A *deterministic* formula is used in all cases to generate the sequence of numbers. The real question you must ask is, "How awful is your random number generator?" Previously I stated that I respect the `drand48()` random number generator and have been using it for years with no apparent problems – I recommend its usage if it is available on your platform.

Most random number generators use the linear congruential algorithm (Morgan 1984).[8] Three constants define the generator, a, c, and m, which cannot be arbitrarily chosen. *Numerical Recipes* (Press *et al.* 1989) provides

[7] This quote, correctly or incorrectly attributed to Einstein, was produced by a computer-login fortune program.

[8] A useful, efficient alternative generates random bits instead of random numbers (Kirkpatrick and Stoll 1981).

a rather large table of possible combinations that work. Given an initial integer, I_0, called the *seed*, the algorithm generates a series of integers, I_1, I_2, I_3, ..., using the equation

$$I_{j+1} = (aI_j + c) \text{ modulus } m, \tag{4.1}$$

where the modulus operator gives the integer remainder after division by m. The "random" integer I_{j+1} provides the seed for the next integer, I_{j+2}, and the random number is obtained by dividing I_{j+1} by m.

It may be helpful to work through the algorithm with the smallest numbers listed in *Numerical Recipes*, $a = 106$, $c = 1283$, and $m = 6075$. Starting with the random seed $I_0 = 4567$ (the seed should be less than m) we examine

$$\begin{aligned}\frac{4567 * 106 + 1283}{6075} &= \frac{484102 + 1283}{6075} \\ &= \frac{485385}{6075} = 79 + \frac{5460}{6075} = 79.898765\ldots\end{aligned}$$

We obtain our next seed from the integer remainder of the division, 5460, and the newly generated random number from the fractional remainder of the division, 0.898765.... We then throw the new seed back into equation (4.1) to get our next iterate

$$\begin{aligned}\frac{5460 * 106 + 1283}{6075} &= \frac{578760 + 1283}{6075} \\ &= \frac{580043}{6075} = 95 + \frac{2918}{6075} = 95.480329\ldots\end{aligned}$$

yielding a new seed, 2918, and random number, 0.480329....

Notice that once the initial seed reoccurs in the random number stream, the entire series of random numbers is repeated. The number of numbers in this series is called the random number generator's cycle length, and in the case of the above algorithm the maximum cycle length is m.

In the program shown below I compare two system-supplied random number generators, `drand48()` and `rand()`, with a basic random number generator from *Numerical Recipes* I call `myran()`. `myran()` and `drand48()` produce random numbers distributed uniformly between the values 0 and 1, and these numbers are used directly to perform the random walk. `rand()` returns a randomly chosen number between 0 and `RAND_MAX` (a constant defined by the compiler) that I change into a uniformly distributed number by dividing by `RAND_MAX`.[9] On many systems, `RAND_MAX`=32,767 ($= 2^{15} - 1$), yet `rand()` has a large cycle length of 2^{32}.

[9] A word of caution with `rand()`: One version of Linux I have has `RAND_MAX`=32,767 ($= 2^{15} - 1$) whereas another has `RAND_MAX`=2,147,483,647 ($= 2^{31} - 1$). In addition the function may actually be used differently (e.g., return a real value, not an integer) on your system.

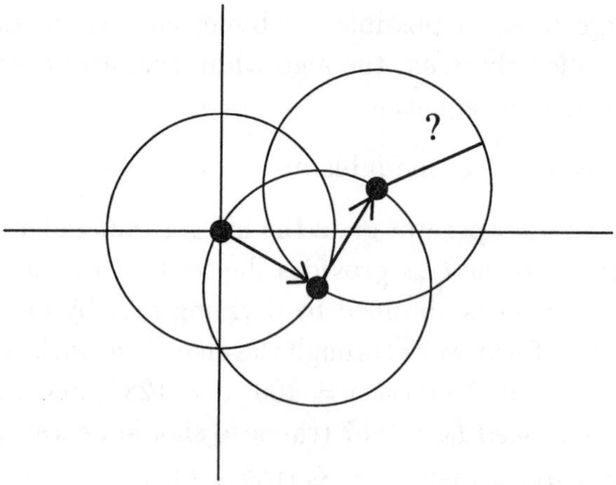

Figure 4.3. The movement rule for the random walker (the filled dot) simulated in the program of section 4.2.2. The walker moves by taking a unit step at a randomly chosen angle.

Although you can write random number generator code explicitly, the complete and utter failure of `myran()` demonstrates why I use `drand48()`. Keep in mind that this poor showing of `myran()` in no way detracts from *Numerical Recipes* as a wonderful resource of numerical algorithms. The authors state "there is a reason that the [simple random number generator] cannot be made as good as the system-supplied" random number generator. `drand48()` uses an algorithm similar to equation (4.1), however system programmers can use very large values for a, c, and m by implementing various tricks,[10] thereby producing good random numbers. All I really want to demonstrate here are the consequences of a really bad random number generator.

4.2.1 Random Walker Test

In the program below I simulate random walkers in continuous, two-dimensional space. There are many ways of performing a random walk; I chose the procedure shown in figure 4.3 in which a walker moves exactly one spatial unit in a randomly chosen direction from its present (x, y) coordinate. Although the algorithm is quite trivial, as shown below it is quite effective in picking out a poor random number generator.

[10] I have no clue as to what those tricks are.

4.2.2 Code

```
     #include  <stdlib.h>
     #include  <stdio.h>
1.a) #include  <math.h>
     #define   SKIPTIME  1000
     #define   STEPS     500
     #define   IC   29573
     #define   IA   3877
     #define   IM   139968

     double    myran(int *);
     void      Step(double *, double *, double);

     int main(void)
     {
          int       i, j, seed1, seed2, seed3;
2.a)      double    x, y, x2, y2, x3, y3;

          seed1 = seed2 = seed3 = 145739853; /* 358937541 */
          srand48(seed1); seed2 = seed2%IM; srand(seed3);

2.b)      x = y = x2 = y2 = x3 = y3 = 0.0;
          printf(" %d  %f  %f  %f  %f  %f  %f \n",
                    0, x, y, x2, y2, x3, y3);

          for(i=0;i<STEPS;i++)
          {
               for(j=0;j<SKIPTIME;j++)
               {
2.c)                Step(&x,&y,drand48());
                    Step(&x2,&y2,myran(&seed2));
                    Step(&x3,&y3,((double)rand()
                         /(double)(RAND_MAX)));
               }
               printf("%d %f %f %f %f %f %f\n",
                    i+1, x, y, x2, y2, x3, y3);
          }
          return (0);
     } /* end main */

2.d) void Step(double *xx, double *yy, double ran)
     {
          double theta;
1.b)      theta = (2.0*M_PI)*ran;
2.e)      *xx += cos(theta); *yy += sin(theta);
     }
```

```
     double myran(int *seed)  /* from Numerical Recipes */
     {
1.f)     *seed = ((*seed)*IA+IC) % IM;
         return((double)(*seed) / (double)IM);
     }
```

4.2.3 Code Details

Line 1: A new header file, `math.h`, is included at line 1.a. This inclusion occurs because in the function `Step()` at line 1.b I use three things defined in the header file. One is the constant, `M_PI`, defined equal to π, and the other two are the declared functions `sin()` and `cos()`. Also, because of these two function calls, a link to the file holding the compiled routines might need to be made when compiling the program. This link is made by the `-lm` flag[11]

```
cc random.c -o brand -lm
```

which links the file `libm.so` or `libm.a` located somewhere in your Unix system's pathways.

Line 2: When variables are defined, for example the variable `x` at line 2.a, the definition tells the compiler to set aside space in memory for the variable. An assignment statement, such as line 2.b, tells the computer to put a specific number into the memory location corresponding to the referenced variable. The address operator, `&`, is used at line 2.c in the argument list of `Step()`. As discussed in chapter 3 on page 59, passing plain old parameters from a calling routine (such as `main()`) to a called routine (such as `Step()`) occurs by copying the parameter list's contents into the local variables defined in the called function. Subsequent changes in local variables within the called routine are not reflected in the variables of the calling routine because the two sets of variables represent completely different locations in memory. Alternatively, as discussed in section 4.1.2, the calling function can pass the address of a variable to the called function. Using this address the called function could access the memory location of the calling function's variable and change its contents. Line 2.d declares `xx` and `yy` as pointers, set to the values of the passed addresses. The memory locations are specified by the calling function (in this case `main()`), but the act of passing the pointers allows the called function to access those

[11] You might not need to include the `-lm` flag with some newer compilers.

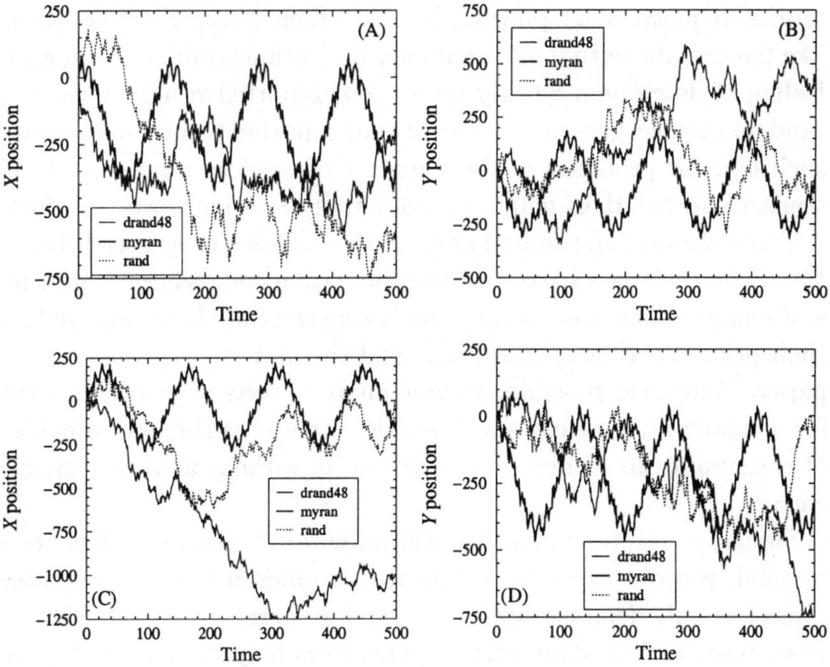

Figure 4.4. Comparisons of three random number generators using the random walk simulation of section 4.2.2. Pairs of panels (A,B) and (C,D) show x and y coordinates as functions of time for two different seeds for the random number generators.

locations. `Step()` does so on line 2.e. Again, why did we use pointers? In essence, we are forced to use pointers when the called function has to return more than one number – in this case `Step()` must return both the new x and y positions.

4.2.4 Random Walker Test Results

Figure 4.4 displays the spatial coordinates for two random walkers as a function of time. Between each plotted step are 1000 unit steps. Everything seems fine except for the cyclic behavior of the x and y positions for the walker that uses `myran()`. This is *very undesirable behavior* for a random number generator, resulting from two problems: A short cycle length and correlations between successive numbers (which we examine below). Imagine your first paper reporting counterintuitive oscillatory population dynamics for some theoretical system whose period is independent of the usual parameters. Three days after being published in *The American Naturalist* the reprint requests start pouring in. You then write a simple little program like the one above that spits out clearly spurious results. With an incredible feeling of doom, you quickly retest your reported results with an improved random number generator to see if your reported results were robust against such horrible problems in the sequence of random numbers. Such possible scenarios, other than making a potentially riveting novel, are just one reason that simulation results in and of themselves are not terribly convincing. Rarely do reviewers see or test the simulation code whose output is reported and analyzed in a manuscript. Reviewers can only base their judgements on their prior experience, knowledge, and the material reported in a submitted paper. Whenever possible, provide analytic tests of your simulation results for as many limiting cases as possible – these benchmarks provide a degree of reassurance to readers that new results without analytic connections are correct.

Similar to the testing of hypotheses within the construct of the scientific method, you can never be certain that a random number generator is good – you can only demonstrate that a random number generator is bad. At the same time, bad random number generators may in one case be acceptable, while being completely unacceptable in other cases. This difference arises because correlations between successive random numbers can be subtle and different correlations may or may not affect the results of a particular simulation. For example, sequential pairs of random numbers may be just fine, but pairs formed from every other random number generated may be correlated. Dependent on the use of the random number generator in your simulation, the results might well be robust against a poor generator. Again, I have not found problems with `drand48()`, hence I feel confident in its use.

4.3 Visualization

Visualization techniques (beyond plotting simple curves of a dependent variable against an independent variable) are an absolute necessity for understanding "nonlinear and complex" systems. Individual-based simulations of ecological systems fall under the rubric of complex systems – systems that have many *degrees of freedom.* In this case each individual is a degree of freedom, or an independent dimension in the phase space of the system. Each individual also has its own set of variables increasing the dimensionality of the system even more. If these complex systems also have nonlinear interactions between the variables, the temporal and spatial dynamics can be very complicated. As variables change via the various model interactions, the system (described by a point in phase space) follows a trajectory through phase space. The overall goal is to understand mechanisms that affect this trajectory.

Given complicated dynamics, how do you comprehend the operative mechanisms and the effects of specific parameters or interactions? One common and useful approach is to first reduce the dimensionality of the system's phase space by replacing many independent variables with a smaller set of composite variables. For example, most population dynamics models use population densities as system variables, rather than a set of variables for each individual. It is much simpler to work with two degrees of freedom, prey and predator densities, rather than thousands of prey and predator variables. Indeed, good, parsimonious models have the fewest degrees of freedom to account for the greatest amount of observed phenomena. After simplifying the model, a theorist then proceeds to examine its dynamics and comprehend the responsible mechanisms, which may assist in further simplifying the model.

This chapter provides visualization techniques allowing maximum exploration of the dynamics possessed by systems with many degrees of freedom thereby revealing mechanistic processes in high-dimension dynamical systems. A major constraint on visualization is the two-dimensional aspect of media such as computer monitors and paper,[12] not to mention the limited dimensional perceptions of the human brain. Often coupled with this limitation is that one of the interesting dimensions in dynamical systems is time. One approach is to use the two dimensions for spatial visualization and perform a temporal animation, but this solution has two drawbacks: the brain cannot process spatial and temporal information simultaneously, and it is difficult to publish animations on the paper of peer-reviewed journals. Another approach sacrifices one spatial dimension for time and uses the remaining dimension for space. The drawback is that a single image contains only one slice of a spatially multidimensional system. Whatever the solution, visualization techniques enable grasping the complicated dynamics produced by these models.

I will introduce PostScript (PS) files within the context of testing random number generators, but subsequent chapters visualize the dynamics of ecological simulations using PS files. PostScript (and Encapsulated PostScript which I use interchangeably) is a language designed by Adobe for printers.[13] On the appropriate printers these files are interpreted as graphics and are output as images. If your printer does not interpret PostScript files, the files

[12] On which you will publish your research work.

[13] Informative websites discussing the use of PostScript programming include those of Lance Lovette and Peter Weingartner.

can be imported into, for example, a spreadsheet and printed from within that software.[14] I use these images in exactly that way – I import them into LaTeXdocuments such as this book. There is some danger that these types of files will become obsolete, however, at the very least, these files can be translated into Adobe's new Portable Document Format (PDF) format.

A common way of testing random number generators is to plot one random number against a previous random number. If the two numbers are completely independent, then there should be no visible patterning in such a plot – it should be random! If, however, a pattern erupts then it demonstrates a correlation between successive random numbers. The following program tests **drand48()**, **myran()**, and **rand()** for such correlations. Given one random number and one from DELAY calls earlier, the element of a two-dimensional array specified by these two integers is incremented. After many such random number pairs are generated and thrown into the array, I produce a square image in which each pixel, corresponding to a specific array element, is blank if no pairs were formed or dark if one or more pairs were formed.

4.3.1 PostScript Visualization Code

```
    #include <stdlib.h>
2.a)#include <stdio.h>

    #define SIDESIZE    (250)
    #define ZOOM        (1.0)
    #define DELAY       (1)
    #define PNTS        (SIDESIZE*SIDESIZE/2/ZOOM/ZOOM)
    #define IC  29573
    #define IA  3877
    #define IM  139968

    double  myran(int *);
    void EPS_Header(FILE *, int, int);
    void EPS_Image(FILE *, int, int, int);
    void EPS_Trailer(FILE *);

1.a)int hits[3][SIDESIZE][SIDESIZE];

    int main(void)
    {
        int     i, j, k, ix, iy, iz, ix2, iy2, iz2;
```

[14] PostScript files can be viewed on a Windows-based PC using the free program available off the web called Ghostscript.

```
        int     seed1, seed2, seed3;
        double  x, y, z;
2.b)    FILE    *fileid1, *fileid2, *fileid3;

2.c)    fileid1 = fopen("image1.eps","w");
        fileid2 = fopen("image2.eps","w");
        fileid3 = fopen("image3.eps","w");

        seed1 = seed2 = seed3 = 145739853; /* 358937541 */
        srand48(seed1);
        seed2 = seed2%IM;
        srand(seed3);

        for(k=0;k<3;k++)
            for(i=0;i<SIDESIZE;i++)
                for(j=0;j<SIDESIZE;j++)
                    hits[k][i][j] = 0;
        x = drand48();
        ix = (int)(SIDESIZE*x/ZOOM);
        y = myran(&seed2);
        iy = (int)(SIDESIZE*y/ZOOM);
        z = (double)rand() / (double)(RAND_MAX);
        iz = (int)(SIDESIZE*z/ZOOM);

        for(i=0;i<PNTS;i++)
        {
            for(j=0;j<DELAY;j++)
            {
                x = drand48();
                y = myran(&seed2);
                z = (double)rand() / (double)(RAND_MAX);
            }
            ix2 = (int)(SIDESIZE*x/ZOOM);
1.b)        if(ix<SIDESIZE && ix2<SIDESIZE) hits[0][ix2][ix]++;
            ix = ix2;
            iy2 = (int)(SIDESIZE*y/ZOOM);
            if(iy<SIDESIZE && iy2<SIDESIZE) hits[1][iy2][iy]++;
            iy = iy2;
            iz2 = (int)(SIDESIZE*z/ZOOM);
            if(iz<SIDESIZE && iz2<SIDESIZE) hits[2][iz2][iz]++;
            iz = iz2;
        }

        EPS_Header(fileid1,SIDESIZE,SIDESIZE);
        EPS_Image(fileid1,0,SIDESIZE,SIDESIZE);
        EPS_Trailer(fileid1);

        EPS_Header(fileid2,SIDESIZE,SIDESIZE);
        EPS_Image(fileid2,1,SIDESIZE,SIDESIZE);
        EPS_Trailer(fileid2);
```

```
    EPS_Header(fileid3,SIDESIZE,SIDESIZE);
    EPS_Image(fileid3,2,SIDESIZE,SIDESIZE);
    EPS_Trailer(fileid3);
    return (0);
} /* end main */

void EPS_Header(FILE *fileid,int width,int length)
{
    double newwid, newlen, rat, xorigin, yorigin;

    newwid = (double)width; newlen = (double)length;
    rat = 500.0/newwid; newwid = 500.0; newlen = rat*newlen;
    if(newlen>700.0)
    {
        rat = 700.0/newlen; newlen = 700.0;
        newwid = rat*newwid;
    }

    xorigin = (612 - newwid)/2.0;
    yorigin = (792 - newlen)/2.0;

2.d)    fprintf(fileid,"%%!PS-Adobe-2.0 EPSF-2.0\n");
    fprintf(fileid,"%%%%BoundingBox: %f %f %f %f\n",
        xorigin-20, yorigin-20,xorigin+newwid,yorigin+newlen);
    fprintf(fileid,"gsave\n");
    fprintf(fileid,"/Times-Roman findfont \n");
    fprintf(fileid,"16 scalefont \n setfont \n");
    fprintf(fileid,"%f %f moveto\n (X) show\n\n",
                    612.0/2.0,yorigin-20);
    fprintf(fileid,"%f %f moveto\n (Y) show\n",
                    xorigin-20,792.0/2.0);
    fprintf(fileid,"%f %f moveto\n",xorigin-1,yorigin-1);
    fprintf(fileid,"%f %f lineto\n",xorigin-1,yorigin+newlen+1);
    fprintf(fileid,"%f %f lineto\n",
                xorigin+newwid+1,yorigin+newlen+1);
    fprintf(fileid,"%f %f lineto\n",xorigin+newwid+1,yorigin-1);
    fprintf(fileid,"%f %f lineto\n",xorigin-1,yorigin-1);
    fprintf(fileid,"stroke\n\n");
    fprintf(fileid,"/bufstr %d string def\n\n",width);
    fprintf(fileid,"%f %f translate\n",xorigin,yorigin);
    fprintf(fileid,"%f %f scale\n\n",newwid,newlen);
    fprintf(fileid,"%d %d 4\n",width,length);
    fprintf(fileid,"[%d 0 0 %d 0 %d]\n",width,-length,length);
    fprintf(fileid,
        "{currentfile bufstr readhexstring pop} bind image\n");
}
```

```
void EPS_Image(FILE *fileid, int mod, int dim1, int dim2)
{
    int i, j, k, flag;

    for(i=dim1-1;i>=0;i--)
    {
        for(j=0;j<dim2;j++)
        {
            if(hits[mod][i][j]==0)
                fprintf(fileid,"F"); /* blank space */
            else
                fprintf(fileid,"0"); /* dark space */
            if((j+1)%60==0) fprintf(fileid,"\n");
        }
    }
    fprintf(fileid,"\n");
}

void EPS_Trailer(FILE *fileid)
{
    fprintf(fileid,"grestore\n");
    fprintf(fileid,"showpage\n");
}
```

4.3.2 Code Details

The `myran()` function on page 85 is unchanged and has not been reproduced above. The commands within the resultant PostScript file are discussed in detail after discussing new C concepts.

4.3.2.1 Multidimensional Arrays.

Line 1: At line 1.a I declare a three-dimensional array. You can think of this array as three separate two-dimensional arrays of size `SIDESIZE`. The program converts random numbers between 0 and 1 into integers between 0 and `SIDESIZE-1`. For each of the three random number generators, given two of these integers, one from the present random number and one from `DELAY` calls earlier, the related array element is incremented at line 1.b. Declaration and use of multidimensional arrays is purely a convenience that high-level programming languages provide – all multidimensional arrays are really just long one-dimensional arrays with an appropriate indexing formula. In this case the three-dimensional array is really an array of length

3*SIDESIZE*SIDESIZE. The order of the array elements in the computer's memory is

```
hits[0][0][0]
hits[0][0][1]
      .
      .
      .
hits[0][0][SIDESIZE-2]
hits[0][0][SIDESIZE-1]
hits[0][1][0]
hits[0][1][1]
      .
      .
      .
hits[0][SIDESIZE-1][SIDESIZE-2]
hits[0][SIDESIZE-1][SIDESIZE-1]
hits[1][0][0]
hits[1][0][1]
      .
      .
      .
hits[2][SIDESIZE-1][SIDESIZE-1].
```

As such, computational efficiency can be gained by working with the physical storage of the array elements and avoiding jumping up and down the array: *As much as possible ensure the rightmost dimensions change most often.*

4.3.2.2 Files.

All of the results from previous chapters' programs were written to the window from which you ran your programs. A straightforward extension to the `printf()` function, the `fprintf()` function, provides output to a file that can subsequently serve as input to another program such as a spreadsheet or a plotting program like `xmgr`. Files are particularly useful if you have programs that take hours or days of CPU time, and are best run as background or batch jobs.[15] A few cautionary remarks. When opening files within your program, take care when opening old files; you might inadvertently overwrite its contents. Also, if your program gets stuck in an infinite loop it may try to write infinite amounts of data, which your computer might not appreciate.

[15] Programs can be run *in the background* on Unix machines by following the compiled program's executable name with a `&`, for example, "`xbd &`". If your run will take many hours, preface this command with `nohup`, for example, "`nohup xbd &`". `nohup` stands for "no hang up" when you log off your interactive session.

Line 2 Line (2.a) includes the `stdio.h` header file that defines various data types associated with file input and output, for example, the variable type `FILE` referenced on line (2.b) (as well as in the function prototypes). Here I define a pointer to a variable of type `FILE` called `fileid1`. These file ids (pronounced ī-'dēz) are used on line (2.c) where the file is actually readied for input/output (io) with a call to the system function `fopen()` that opens files, making them ready for input or output. Lots of operating system stuff goes on behind the scene during one of these calls, which I treat as a black box. There are two arguments for `fopen()`, the first being the name of the opened file contained within double-quotation marks. If the file already exists, it is connected to the program; if it does not exist, it is created and then connected to the program. The second parameter, `"w"`, defines the operation that can be performed on the file. The list of allowable file operations include `"r"` and `"w"`, for read and write. Clearly, if you are going to read from a file, it must first exist – if the file does not exist the function returns an error flag.[16] Both parameters, the filename and mode, are placed within double quotes to signify that they are strings of text. Finally, line (2.d) demonstrates writing to a file with the `fprintf` function. This function works just like the `printf` function except that the first argument is a pointer to a file. When the program finishes you will find a file in the directory from which you ran your program with the name `image1.ps`. You could open up this file and edit its contents with your favorite editor if you wanted to.

There are lots of options when working with files, including when and how data are stored. Three useful functions are `fflush()` which flushes the output buffer of data written to a file, `fclose()` which closes a file during program execution, and the function `fseek()` which places the input/output pointer to a specific location. Above, I showed you how to store data as text files because they are easily imported into data processing programs. A more efficient method of storing mountains of data is in binary files. These files aren't viewable by "pesky humans," rather they hold the data exactly like the computer uses it. For example, I often store a complete equilibrium state of an individual-based simulation in a file readable into subsequent simulation runs. Functions for binary input/output are `fwrite()` and `fread()` with usages very different from the above functions; they are the best options for binary file generation. These functions directly copy the contents from a specific number of bytes in file storage to a region of memory iden-

[16] Professional programmers would always test for these errors; I never check because figuring out why my program crashed in these rare cases is less work than writing code for all the possible checks.

tified by a pointer, or vice versa. Binary files are preferable under many conditions, particularly storing large arrays or structures, and it might be well worth your while to learn about them.

4.3.2.3 PostScript file.

I am skipping coverage of spreadsheets and plotting programs, like the Unix plotting program **xmgr**, that plot curves of a dependent variable against an independent variable. These are important visualization methods, but these change too rapidly,[17] making any discussion obsolete rather quickly. I am therefore moving on to visualization methods with maximum programmer control. Below I reproduced, and discuss line-by-line, the PostScript file resulting from the three calls to the **EPS** functions.

```
1)    %!PS-Adobe-2.0 EPSF-2.0
2)    %%BoundingBox: 161.000000 251.000000 431.000000 521.000000
3.a) gsave
4.a) /Times-Roman findfont
4.b) 16 scalefont
4.c) setfont
4.d) 306.000000 251.000000 moveto
4.e) (X) show
     161.000000 396.000000 moveto
     (Y) show

5.a) 180.000000 270.000000 moveto
5.b) 180.000000 522.000000 lineto
     432.000000 522.000000 lineto
     432.000000 270.000000 lineto
     180.000000 270.000000 lineto
5.c) stroke

6.a) /bufstr 250 string def

6.b) 181.000000 271.000000 translate
6.c) 250.000000 250.000000 scale

6.d) 250 250 4
6.e) [250 0 0 -250 0 250]
6.f) {currentfile bufstr readhexstring pop} bind image
6.g) F0FFF00FFFFF00F000FFF0FFFFFFF0FF0FFF0F0FF0FF0F0F0F00FFFF0F0F
     FFFFF0FFFFFFF000FFFFFFFFF0FFFFFF0F0FFFFF0F0FFF0FFFFFF0F0FF00
     FF0F0F00F000F00F0FFFF000FFFFFFF0FFFFFF0FFF0FFF0F000F00FF0FF0
                                 :
     F0FFF000F00F0FFFF0FF000FFFFFFFFF0FFFF0F0F0F0FF00FFFFFFF0F0FF0
```

[17] For example, **xmgr** is being replaced by **grace**.

```
       F0000F0F0F0F0FFF0F0FFFF000FFF00FF000FFFFFFFFF00FFF0FF00F00FF
       F0F0FFFF0F
3.b) grestore
7)     showpage
```

Line 1: This line defines the file type as a PostScript file – all operating systems look for the first two characters.

Line 2: This line defines the size of the image. An 8.5x11 page, with no margins, is 612x792 points in size. The PostScript page is laid out with the origin `(0,0)` at the lower left and the point `(612,792)` at the top right. Devices or software packages that deal with Encapsulated PostScript files need this information.

Line 3: At line 3.a the present graphics state is saved. This means that the PostScript file can do whatever it wants, then restore the previous graphics state at line 3.b. Commercial programs written by professional programmers get into trouble if this save and restore is not performed correctly.

Line 4: This set of lines demonstrates placing text within the PostScript image, although I rarely use this feature – it is much easier to place objects such as text and arrows into an image from within canned graphics software. At line 4.a the desired font is specified and found, its size specified at (4.b), then set at line 4.c. Line 4.d places the cursor at a specific location on this page. Finally, an **X** is written at the present location on line 4.e.

Line 5: This collection of lines draws a box around the image. Line 5.a sets the cursor at a specific spot, then line 5.b marks a line from the present position to the specified position and places the cursor at the new position. Four lines are marked, then at line 5.c these marked lines are filled in by the `stroke` command.

Line 6: Line 6.a defines a new command, `bufstr`, that sets up strings of the specified size at line 6.f. Line 6.b moves the origin to the specified location, then scales the to-be-read image to the specified size at (6.c). The first two items on line 6.d specify the dimensions of the image in terms of data points, and the final item specifies that each point is a 4-bit datum. A datum can consist of 1, 2, 4, 8 or 12 bits. Line 6.e defines matrix parameters telling the image command that the data are arranged left to right and top to bottom. Line 6.f is a procedure that reads the data starting on line 6.g. The data are written in hexadecimal notation (see page 77), where F refers to a blank pixel, 0 refers to a black pixel, and numbers in between refer to various shades of grey.

Line 7: This command tells the printer to print or the equivalent action for a graphics program. I have read that the command can cause problems for various programs, and you may have to remove it.

4.3.3 Visualization Results

The PostScript images in figure 4.5 from `drand48()` and `rand()` look pretty random (there exist more rigorous statistical tests for randomness), but the lower three images from `myran()` have regions with vaguely striated patterns. It doesn't matter what the pattern is – anything other than a random pattern is undesirable, hence, `myran()` fails.

Images in figure 4.6 zoom in on the lower left 0.1×0.1 region of the images shown in figure 4.5 and increase the length of the run by setting `ZOOM` equal to 0.1. These images show that even the best-case scenario for `myran()` has a degree of fine-grained structure that could cause problems. In actuality, all random number generators will give a similar picture after zooming in enough, simply because the floating-point numbers are represented by a finite number of bits.

4.4 Nonuniformly Distributed Random Numbers

So far all the random numbers we used were distributed uniformly between 0 and 1. These random numbers are called *uniform deviates*, because the probability distribution for the numbers is constant, or uniform, within some range. Often random numbers distributed according to nonuniform probability distributions, for example, Poisson or normal distributions, are needed to model well-known underlying biological or physical processes (Morgan 1984). In this section I will introduce the basic guidelines for obtaining nonuniform deviates using the random numbers produced by `drand48()`. The ideas presented here are described in detail in *Numerical Recipes.*

Assume you have a probability distribution $p(x)$ (see figure 4.7) that defines the probability of obtaining the value x. All probability distributions are normalized to one over the relevant interval, such that

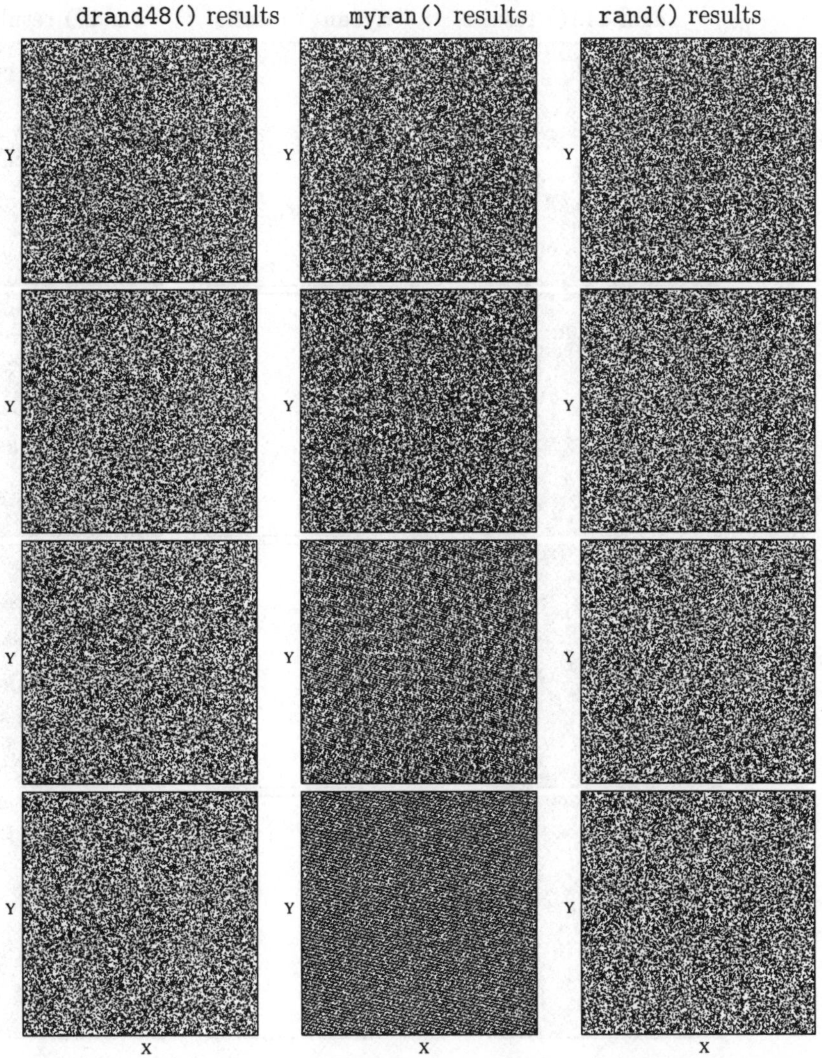

Figure 4.5. PostScript images plotting one random number against a prior random number. The first column uses `drand48()`, the second column uses `myran()`, and the last column uses `rand()`. The nth row plots x_{i+n} vs x_i. Although testing whether a pattern is truly random is hard, the patterns in the middle column are readily observable. These patterns indicate that the random number generator `myran()` is bad!

$$1 = \int_a^b p(x)dx, \tag{4.2}$$

where a represents the lower bound of the distribution and b the upper

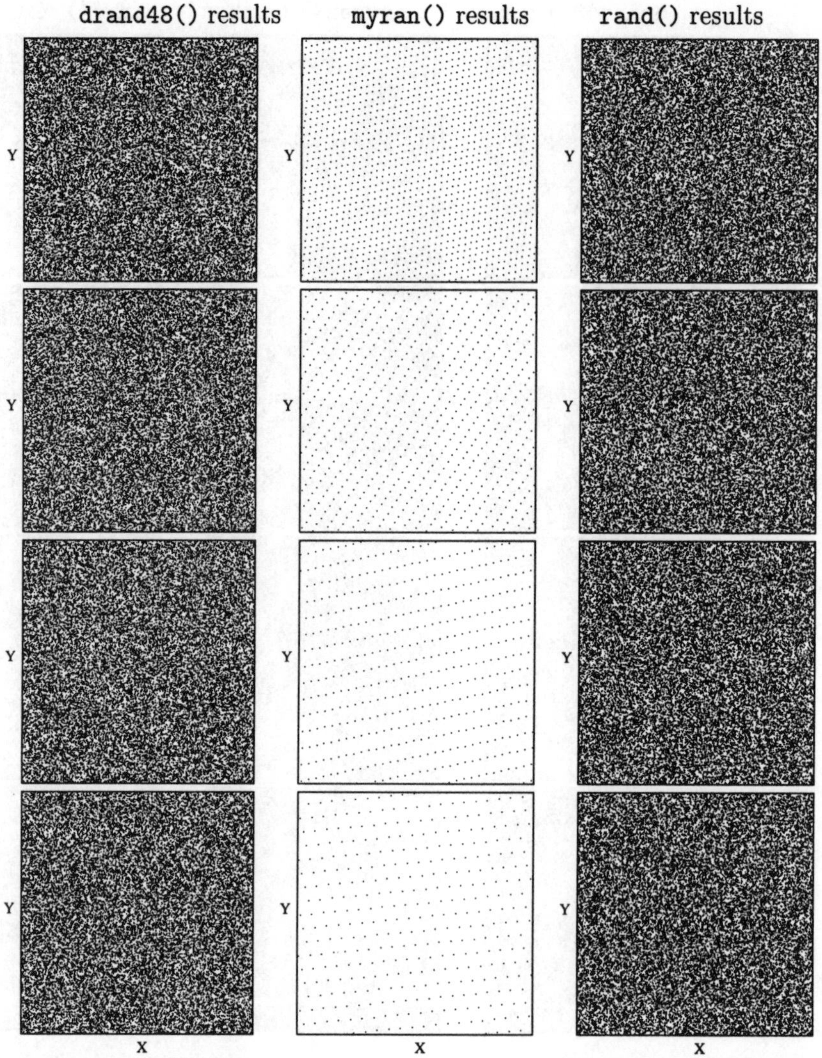

Figure 4.6. PostScript images similar to those in figure 4.5, except zooming in on the region where both numbers are less than 0.1. Rows and columns are arranged identically to figure 4.5. Again, emphatically, `myran()` is bad!

bound. The integral of the distribution up to the point y, $f(y)$, equals unity at the upper bound. Define a new variable, z, equal to the function $f(y)$, taking on values between zero and one, $0 \le z \le 1$. At this point we have

$$z = f(y) = \int_a^y p(x)dx. \tag{4.3}$$

This expression relates a uniform deviate z (e.g., from **`drand48()`**) to a

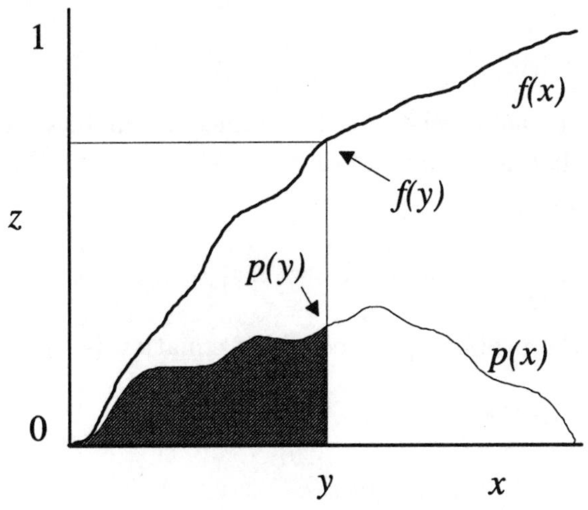

Figure 4.7. Transformation between uniform and nonuniform deviates (after *Numerical Recipes*). The thin line represents a probability distribution with a unit area under the curve. Integrating up to a point y yields a number $z = f(y)$ between 0 and 1, just like random number generators. We can exploit the relationship between z and $p(x)$ to generate arbitrarily distributed random numbers.

value for y – a random number distributed according to the distribution p. Formally, this relation is obtained by differentiating the left and right sides of equation (4.3) with respect to y to obtain

$$\frac{dz}{dy} = p(y). \tag{4.4}$$

Equation (4.4) is an ordinary differential equation; the expression we are looking for is obtained by solving equation (4.4) and inverting its solution to obtain the function $y(z)$, i.e., the random number y (which is distributed according to p) as a function of the uniform variate, z, that the computer gives us.

4.4.1 Exponentially Distributed Random Numbers

Many processes in the biological and physical sciences are described by exponential distributions. Any process in which events occur with constant probability, such as simple models of mortality and the decay of radioactive nuclei, produce a distribution of event times distributed as an exponential distribution

$$p(x) = Ae^{-\alpha x}, \tag{4.5}$$

where A is the as-of-yet-to-be-determined normalization constant assuring that the probability distribution integrates to one. We determine A from equation (4.2)

$$1 = A\int_0^\infty e^{-\alpha x}dx = A\frac{1}{\alpha} \Rightarrow A = \alpha. \tag{4.6}$$

The rest of the problem proceeds from equation (4.4), given in this case as

$$\frac{dz}{dy} = \alpha e^{-\alpha y}. \tag{4.7}$$

Integrating equation (4.7) gives

$$z(y) = -e^{-\alpha y} + C. \tag{4.8}$$

The integration constant, C, is determined by realizing that the lower bound of the probability distribution maps to $z=0$, or $z(y=0)=0$. This condition yields $C = 1$. Then we have

$$\begin{aligned} z &= 1 - e^{-\alpha y} \\ e^{-\alpha y} &= 1 - z \\ -\alpha y &= \ln(1 - z). \end{aligned} \tag{4.9}$$

If z is a uniform deviate then so is $1 - z$, which itself can be replaced by the variable name z. Finally, we get the exponentially distributed deviate y in terms of the uniformly distributed deviate z

$$y = -\frac{1}{\alpha}\ln z. \tag{4.10}$$

As you might guess, this process becomes more difficult when the distributions become more complicated. However, even in the worst-case scenario, equation (4.3) at least provides a numerical approach to determining nonuniform deviates.

4.4.2 Normally Distributed Random Numbers

Many biological and physical quantities are distributed according to a "normal" (also called Gaussian) distribution

$$p(x) = \frac{1}{\sqrt{2\pi}\,\sigma}\,e^{-\frac{x^2}{2\sigma^2}}, \tag{4.11}$$

where zero is the average value for x and σ is the standard deviation. The Box-Muller formula (Box and Muller 1958, Morgan 1984) for obtaining normally distributed deviates transforms two uniformly distributed deviates, z_1 and z_2, into two normally distributed deviates, y_1 and y_2. The expressions for the transformation are

$$y_1 = \sqrt{-2\ \ln z_1}\ \sigma \sin(2\pi z_2) \quad (4.12)$$
$$y_2 = \sqrt{-2\ \ln z_1}\ \sigma \cos(2\pi z_2) \quad (4.13)$$

(the mathematical details for solving this example can be found in *Numerical Recipes*).

Here are four lines of C code I use in an individual-based simulation that incorporates long-range dispersal. An organism's new location is distributed normally about its present location with a standard deviation equal to `displen`. The code uses two calls to `drand48()`, and provides the final positions for the organism in `newrow` and `newcol`

```
sqroot = sqrt(-2.0*log(drand48()));
theta = 6.2831853*drand48();
newrow = oldrow + (int)(sqroot*sin(theta)*displen)
newcol = oldcol + (int)(sqroot*cos(theta)*displen)
```

4.5 Exercises

4.1 Let's play with your computer's memory. Run the following code and describe what happens.

```
int main(void)
{
        int i, array[100];
        for(i=0;i<100;i++) array[i] = 0;
        for(i=0;i<1000;i++)
            if(i%10==0) printf(" %d %d \n",i, array[i]);
        return (0);
}
```

4.2 Mullish and Cooper in *The spirit of C* suggest a random number generator based on the update algorithm

```
seed=((seed*seed)/(long)100)%(long)10000;
```

The variable `seed` is of type `long` and serves as both the random number distributed between 0000 and 9999, and the seed for the next iterate. Something bad happens when `seed=0000` – what? Devise a way to fix that problem, then examine the PostScript plots of this random number generator. Is it OK? If not, can you make it better?

4.3 Convert the logistic birth–death simulation to use the exponentially distributed random numbers discussed above. This conversion requires two parts. First, the net rate of something happening is the sum of the birth and death rates, and this net rate defines the probability distribution for event times, $P(\tau) = P_0 \exp(-(\alpha(N)+\beta(N))\tau)$. Hence, one random number is used to find the time when the next event occurs. Second, is the event a birth or a death? Answering this question requires another random number to check against the probability that the event is, say, a birth, $\alpha(N)/(\alpha(N)+\beta(N))$. This revised code should give a much more efficient simulation (fewer random numbers per realized event) and better agreement with the analytic predictions. Is it more efficient and does it provide better agreement?

4.4 See how the terminal screen is treated exactly like a file. In the visualization program, comment out the `fileid1 = fopen( ... )` line and replace `fileid1` with `stdout` everywhere else. Compile and run your program – what happens? [Be sure to include `stdio.h`!]

4.5 Write a PostScript-producing routine to plot the two-dimensional path of a random walker.

4.6 Get the visualization code working, including generation of the PostScript images, and compare `myran()` images for delays of 2^n and $2^n + 1$ using values of $n = 3$, 4, 5, and 6.

5

Two-Species Competition Model

Programming components covered in this chapter:

- conditional compilation
- more arrays and functions
- using PostScript files

There are two fundamentally different ways that organisms compete, directly and indirectly. A direct form of competition, called interference competition, is a one-on-one battle resulting, in the extreme case, in one of the competitors perishing. Interference competition removes individuals in proportion to the product of the competing species' densities. One indirect form is competition for common resources, called exploitation competition. The organism using the resource first wins it, thereby making the resource unavailable to others. Hence, exploitation competition by one species reduces the growth rate of competitors in proportion to the first species's population density.[1]

An important mathematical formulation of competition is the two-species Lotka–Volterra competion model (e.g., Case 2000) studied in this chapter. Two very important parameters determining the outcome of competition in this model are the *competition coefficients* describing the effect one species has on the other. In this chapter, I put together a simulation model combining aspects of exploitation and interference competition that translates directly into the framework of a Lotka–Volterra competition model. The model's competition coefficients turn out to be aggregates of the arguably more mechanistic simulation model's parameters, as they were for the single-species logistic growth model.

[1] There are complicated competitive situations. For example, organisms competing exploitatively for space, but the competition is unapparent until space is almost fully occupied, at which point there are one-on-one battles for the few remaining spatial locations has been called "preemptive competition" (Morin 1999).

We will examine both nonspatial and spatial versions of the simulation and find that in the spatial version another aspect of competition arises making individual dispersal an important factor in competitive outcomes (Levin, Cohen, and Hastings 1984, Vance 1985). We also will explore the importance of dispersal using a crude analysis of a deterministic spatial Lotka–Volterra competition model.

5.1 Analytic Lotka–Volterra Competition Model

5.1.1 Model Interactions

Imagine a system of two competing species with population densities N_1 and N_2. Each species has a species-specific reproduction rate α_i and a mortality rate μ_i, where $i = 1, 2$ for the two competitors. Assume both species have the same carrying capacity K set by exploitation competition for space. Imagine limpets on a rock or plants on a hillside – only one organism can be at one location at any instant. In addition, there are two interference competition parameters β_{12} and β_{21} specifying the mortality imposed on species i by interactions with species j. We can translate these interactions into two coupled differential equations

$$\frac{dN_1}{dt} = \alpha_1 N_1 \left(1 - \frac{N_1}{K} - \frac{N_2}{K}\right) - 2\beta_{12} N_1 N_2 - \mu_1 N_1 \tag{5.1a}$$

$$\frac{dN_2}{dt} = \alpha_2 N_2 \left(1 - \frac{N_1}{K} - \frac{N_2}{K}\right) - 2\beta_{21} N_1 N_2 - \mu_2 N_2. \tag{5.1b}$$

In both expressions, the first term has a factor $(1-N_1/K-N_2/K)$, representing the probability that any randomly chosen site is devoid of an organism. These empty sites are the only ones available for colonization by offspring of either species. One important assumption about competition contained here is that members of both species can hang on to their sites, resisting invasion by juveniles of the competing species. The other factor in the first terms, for example $\alpha_1 N_1$, represents the offspring production rate by all adults of the species. In summary, the first term represents the rate that offspring are both produced *and* subsequently placed in empty sites.

The second term in both equations has a product $N_1 N_2$. Imagine a pair of neighboring sites chosen randomly from the habitat: What is the probability that one site is occupied by an individual of species 1 and the other site is occupied by an individual of species 2? Assuming sites are occupied independently, the dual occupancy probability is just the product of the two probabilities, multiplied by a factor of 2 because there are two ways of placing two distinct individuals in two different boxes. The third term is easy – it represents the proportion of individuals that perish per unit time.

Rearranging the terms in equation (5.1) yields

$$\frac{dN_1}{dt} = \frac{\alpha_1 N_1}{K}\left(\frac{(\alpha_1 - \mu_1)K}{\alpha_1} - N_1 - \frac{(2\beta_{12}K + \alpha_1)}{\alpha_1}N_2\right), \quad (5.2a)$$

$$\frac{dN_2}{dt} = \frac{\alpha_2 N_2}{K}\left(\frac{(\alpha_2 - \mu_2)K}{\alpha_2} - \frac{(2\beta_{21}K + \alpha_2)}{\alpha_2}N_1 - N_2\right). \quad (5.2b)$$

After defining a new set of parameters, this set of equations matches the form of the classical Lotka–Volterra competition model (e.g., Hastings 1997)

$$\frac{dN_1}{dt} = \frac{r_1 N_1}{K_1}(K_1 - N_1 - a_{12}N_2) \quad (5.3a)$$

$$\frac{dN_2}{dt} = \frac{r_2 N_2}{K_2}(K_2 - a_{21}N_1 - N_2), \quad (5.3b)$$

where the new parameters are defined in terms of the original model's parameters

$$r_1 = \alpha_1 - \mu_1, \qquad r_2 = \alpha_2 - \mu_2,$$

$$K_1 = \frac{\alpha_1 - \mu_1}{\alpha_1}K, \qquad K_2 = \frac{\alpha_2 - \mu_2}{\alpha_2}K,$$

$$a_{12} = \frac{2\beta_{12}K + \alpha_1}{\alpha_1}, \qquad a_{21} = \frac{2\beta_{21}K + \alpha_2}{\alpha_2}.$$

α_1 and α_2 are the competition coefficients which, in this example, are phenomenological parameters made up of several more mechanistic parameters. Hence, in a similar way that the logistic model can represent many different sets of assumptions regarding single-species birth and death functions (see chapter 3), the Lotka–Volterra competition model is conceptually vague, covering many sets of assumptions for specific competitive interactions. Criticisms have also been leveled at the Lotka–Volterra competition model for important biological reasons ranging from time delays (Caswell 1972) to mechanisms (Comins and Hassell 1987, Tilman 1987).

5.1.2 Coexistence Conditions

Here we derive the conditions needed for the two species to coexist. This derivation will be a two-part approach closely following that of MacArthur and Levins (1967) examining mutual invasibility of single-species systems by the other species. First, in systems containing only conspecifics the

Table 5.1. Connections between original and modified competition models

Original model		Modified model		
Parameter	Value	Parameter	Expression	Value
α_1	0.25	r_1	$\alpha_1 - \mu_1$	0.2
α_2	0.25	r_2	$\alpha_2 - \mu_2$	0.2
β_{12}	0.075	a_{12}	$\frac{2\beta_{12}K+\alpha_1}{\alpha_1}$	1.6
β_{21}	0.1	a_{21}	$\frac{2\beta_{21}K+\alpha_2}{\alpha_2}$	1.8
μ_1	0.05	K_1	$\frac{\alpha_1-\mu_1}{\alpha_1}K$	0.8
μ_2	0.05	K_2	$\frac{\alpha_2-\mu_2}{\alpha_2}K$	0.8
K	1.0			

equations (5.1a) and (5.1b) collapse to

$$\frac{dN_1}{dt} = \frac{r_1 N_1}{K_1}(K_1 - N_1) \tag{5.4a}$$

$$\frac{dN_2}{dt} = \frac{r_2 N_2}{K_2}(K_2 - N_2). \tag{5.4b}$$

The equilibrium species densities (specified by $dN_1/dt = dN_2/dt = 0$) are the effective carrying capacities K_i. Second, in systems dominated by heterospecifics (for example when N_1 is small compared to K_1 and $a_{12}N_2$), the equations collapse instead to

$$\frac{dN_1}{dt} = \frac{r_1 N_1}{K_1}(K_1 - a_{12}N_2) \tag{5.5a}$$

$$\frac{dN_2}{dt} = \frac{r_2 N_2}{K_2}(K_2 - a_{21}N_1). \tag{5.5b}$$

One reasonable definition of coexistence is that both species have positive growth rates when placed in a system with the other species at its equilibrium density. These conditions yield $K_1 > a_{12}K_2$ and $K_2 > a_{21}K_1$, leading to the usual Lotka–Volterra condition for coexistence

$$a_{12} < \frac{K_1}{K_2} < \frac{1}{a_{21}}, \tag{5.6}$$

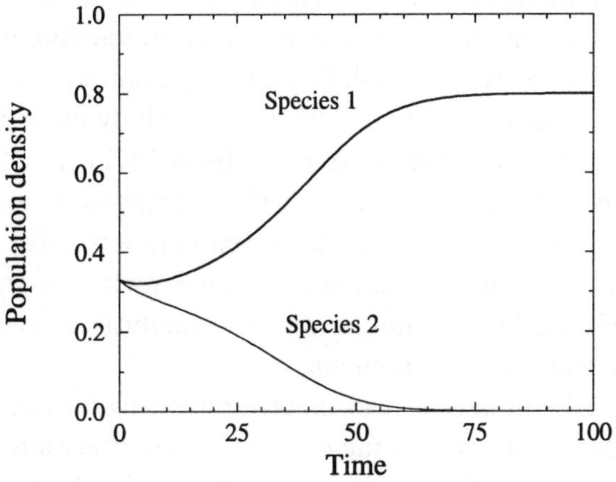

Figure 5.1. Population dynamics of the analytic competition model, equations (5.3), using the parameters listed in table 5.1. Initial density for both species is $N_1(0) = N_2(0) = 1/3$.

or in terms of our original parameters

$$\frac{2K\beta_{12} + \alpha_1}{\alpha_1} < \frac{\frac{\alpha_1 - \mu_1}{\alpha_1}K}{\frac{\alpha_2 - \mu_2}{\alpha_2}K} < \frac{1}{\frac{2K\beta_{21}+\alpha_2}{\alpha_2}} \tag{5.7}$$

$$\frac{2K\beta_{12} + \alpha_1}{\alpha_2} < \frac{\alpha_1 - \mu_1}{\alpha_2 - \mu_2} < \frac{\alpha_1}{2K\beta_{21} + \alpha_2}. \tag{5.8}$$

For the sake of a specific example to use throughout this chapter, let's use the values listed in table 5.1 for the original model and the resultant values for the "transformed" Lotka–Volterra model. I chose values such that both species are identical in all respects, except that species 1 is a better competitor than species 2 with respect to the β_{ij} parameters.

Replacing the transformed Lotka–Volterra parameters into the coexistence condition, we have

$$1.6 < \frac{0.8}{0.8} < \frac{1}{1.8}. \tag{5.9}$$

Both conditions are violated, meaning that there is *no coexistence* and the winning species depends on the initial condition. A simple numerical integration (see exercise 5.4) of the original model using the parameter values listed in table 5.1 confirms the lack of coexistence. The dynamics are shown in figure 5.1 for initial densities $N_1(0) = N_2(0) = 1/3$.

5.2 Simulation Model of Lotka–Volterra Interactions

Building a simulation model of the above competitive system will test the analytic predictions from an individual-based viewpoint. There will be one major difference between the differential equation model discussed above and the simulation model: Space. Although spatially explicit stochastic models can be examined analytically (see Mollison 1977), much work is limited to models of cell dynamics (Zeigler 1977). Important examples include the model by Silvertown *et al.* (1992) which competed five species against one another in a two-dimensional habitat, and Schwinning and Parsons's (1996a, 1996b) work combining simulations and analytic models to provide useful insight into empirical phenomena.

This spatial simulation also incorporates several new assumptions, each of which potentially affects the outcome of competition. One assumption is that space consists of a one-dimensional line of cells, with each cell containing at most one individual, allowing us some hope for analytic connections (Allen, Allen, and Gilliam 1996, Hart and Gardner 1997). Also, an offspring can only be placed into an empty cell, and once there it never leaves the cell until it dies. The next assumption is the interesting one: Offspring are placed within a species-specific distance from their parent's cell. To make things really interesting later, we will assume that the poorer competitor, species 2, has the longer offspring placement distance. This set of assumptions corresponds with the dispersal–competitive ability tradeoff observed in several biological systems (see Holmes and Wilson 1998).

5.2.1 Code

```
#include <stdlib.h>
#include <stdio.h>

#define   MAXTIME          100
#define   OUTTIME          5
#define   PSTIME           1
#define   CELLS            1000
#define   PSSKIP           2
#define   ALPHA1           0.25
#define   ALPHA2           0.25
#define   BETA12           0.075 /* b12+b21<0 */
#define   BETA21           0.1
#define   MU1        0.05
#define   MU2        0.05
#define   MIXED            1
```

```
int       state[CELLS], dist1, dist2;
double    n1cnt, n2cnt, alpha1, alpha2,
          beta12, beta21, mu1, mu2;

void InitStuff(void);
void Competition(void);
void Reproduction(void);
void NewBabe(int, int, int);
void Mortality(void);
#if(MIXED)
void Mix(void);
#endif
void Measurements(void);
void EPS_Header(FILE *, int, int);
void EPS_Row(FILE *, int);
void EPS_Trailer(FILE *);

int main(void)
{
     int i;
     FILE *fileid1, *fileid2;

     fileid1 = fopen("compete1.eps","w");
     fileid2 = fopen("compete2.eps","w");
     InitStuff();
     printf("Enter dist1 and dist2:");
     scanf("%d %d",&dist1,&dist2);

     EPS_Header(fileid1,CELLS/PSSKIP,MAXTIME/PSTIME);
     EPS_Header(fileid2,CELLS/PSSKIP,MAXTIME/PSTIME);

     for(i=0;i<MAXTIME;i++)
     {
          Measurements();
          if(i%PSTIME==0)
          {
               EPS_Row(fileid1,1);
               EPS_Row(fileid2,2);
          }
          if(i%OUTTIME==0) printf("%d %f %f\n",
                            i, n1cnt, n2cnt);

          Mortality();
          Competition();
          Reproduction();
#if(MIXED)
2.a)      Mix();
#endif
     }
     EPS_Trailer(fileid1); EPS_Trailer(fileid2);
```

```
        return (0);
    }

    void InitStuff(void)
    {
        int i, seed=145739853;
        double xxx;

        srand48(seed);
        alpha1 = ALPHA1; alpha2 = ALPHA2;
        beta12 = BETA12; beta21 = BETA21;
        mu1 = MU1;       mu2 = MU2;

        for(i=0;i<CELLS;i++)
        {
            xxx = drand48();
            if(xxx<0.33)      state[i] = 1;
            else if(xxx<0.66) state[i] = 2;
            else              state[i] = 0;
        }
    }

    void Competition(void)
    {
        int i, j, k, pairs[CELLS], npair, totpairs;
        double xxx;

        totpairs = 0;
        for(i=0;i<CELLS;i++)
1)          if(state[i]*state[i+1]==2)
                pairs[totpairs++] = i;

        k = totpairs;
        for(i=0;i<totpairs;i++)
        {
            j = k*drand48();
            npair = pairs[j];
            if(state[npair]*state[npair+1]==2)
            {
                xxx = drand48();
                if(xxx<beta12)
                {
                    if(state[npair]==1) state[npair] = 0;
                    else state[npair+1] = 0;
                }
                else if(xxx<beta12+beta21)
                {
                    if(state[npair]==2) state[npair] = 0;
                    else state[npair+1] = 0;
                }
```

```
        }
        pairs[j] = pairs[--k];
    }
}

void Reproduction(void)
{
    int i, j, k, sites[CELLS], nsite, totpop=0;

    for(i=0;i<CELLS;i++)
    {
        if(state[i]==1)
        {
            if(drand48()<alpha1)
                sites[totpop++] = i;
        }
        else if(state[i]==2)
        {
            if(drand48()<alpha2)
                sites[totpop++] = i;
        }
    }

    k = totpop;
    for(i=0;i<totpop;i++)
    {
        j = k*drand48();
        nsite = sites[j];
        if(state[nsite]==1) NewBabe(nsite,1,dist1);
        else                NewBabe(nsite,2,dist2);
        sites[j] = sites[--k];
    }
}

void NewBabe(int isite, int species, int dist)
{
    int newsite;

    newsite = dist*drand48() + 1;
    if(drand48()<0.5) newsite = -newsite;
    newsite += isite;
    if(newsite>=0&&newsite<CELLS)
        if(state[newsite]==0) state[newsite] = species;
}

void Mortality(void)
{
    int i;
    double xxx;
```

```
        for(i=0;i<CELLS;i++)
        {
            xxx = drand48();
            if(state[i]==1 && xxx<mu1) state[i] = 0;
            if(state[i]==2 && xxx<mu2) state[i] = 0;
        }
    }

    #if(MIXED)
2.b) void Mix(void)
    {
        int i, j, k, temp;
        double xxx;

        for(i=0;i<CELLS;i++)
        {
            j = CELLS*drand48(); k = CELLS*drand48();
            temp = state[j];
            state[j] = state[k];
            state[k] = temp;
        }
    }
    #endif

    void Measurements(void)
    {
        int i;

        n1cnt = n2cnt = 0.0;
        for(i=0;i<CELLS;i++)
        {
            if(state[i]==1)          n1cnt += 1.0;
            else if(state[i]==2)     n2cnt += 1.0;
        }
        n1cnt /= (double)(CELLS); n2cnt /= (double)(CELLS);
    }

    void EPS_Row(FILE *fileid, int species)
    {
        int  i;

        for(i=0;i<CELLS;i+=PSSKIP)
        {
            if(state[i]==species)    fprintf(fileid,"0");
            else                     fprintf(fileid,"F");
            if((j+1)%60==0) fprintf(fileid,"\n");
        }
        fprintf(fileid,"\n");
    }
```

5.2.2 Code Details

Line 1: This line tests whether both of a pair of neighboring sites are occupied by neighbors of different species. If either site is unoccupied, then the product is zero. If the sites are occupied by the same type, then the product is either 1 or 4. Only if the product is 2 is the pair marked for additional tests. The primary reason I perform this initial testing is to set up a list of paired sites from which I can choose pairs in a random order rather than sweeping across the lattice in one direction or another. Such sweeps can set up bad correlations (see exercise 5.6).

Line 2: A major role is played by the spatial structure that develops in the distribution of the two species, and this structure removes congruence with the nonspatial model equations. On line 2.a I call a routine that homogeneously mixes the populations, `Mix()` (found at line 2.b), removing all spatial structure by swapping the states of randomly chosen cells. I choose `CELLS` pairs of cells, meaning that each cell is swapped with another an average of twice per time step. Likely, this much swapping is overkill!

5.2.3 Nonspatial Simulation Results

Our main comparison between the simulation results and the predictions of the analytic model will be for the mixed case which eliminates the effects of space, leaving only the effects of stochasticity. Figure 5.2A compares three replicate simulation runs with the deterministic results shown in figure 5.1. The results of the simulation roughly match the expectations of the analytic model.

Figure 5.3 are space–time pictures of the lattice distributions for both species for a single run. The horizontal dimension shows the spatial axis, and the time runs downward. We see species 1 reaching a high density and species 2 being excluded. All of the information shown in these plots can be gleaned from the plot of population counts, but I included these plots to demonstrate the effective removal of all spatial correlations by using the mixing algorithm. Although a thorough mathematical analysis of a stochastic version of the competition model, along the lines of the analysis performed for the logistic single-species model has been done (Leslie 1958, Goel and Richter-Dyn 1974), we will not cover it here.

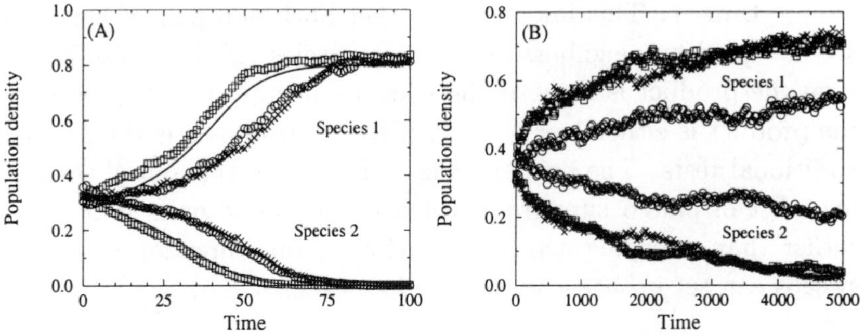

Figure 5.2. (A) Numerical comparison of analytic results (figure 5.1) with several spatially mixed simulation runs. (B) Population dynamics for several runs of the spatial competition simulation with `dist1=dist2=1`. As in the mixed case, species 1 wins but its takeover time is drastically increased.

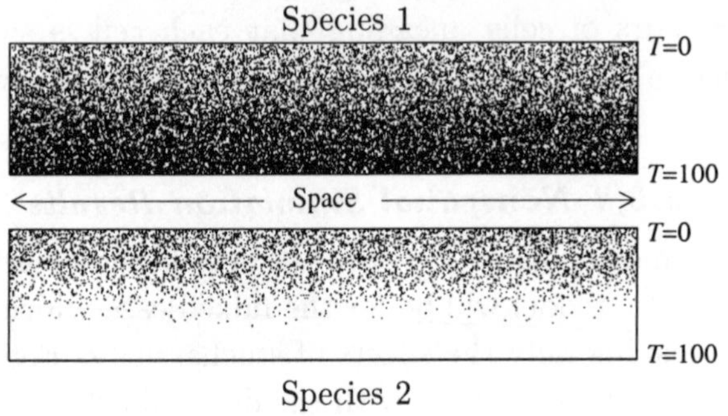

Figure 5.3. PostScript files from one spatially mixed simulation run showing the lack of spatial structure, and hence the effectiveness of the mixing algorithm, in the two populations. Dark pixels indicate individuals, thus species 1 wins over species 2.

5.2.4 Spatial Simulation Results

Changing the `MIXED` flag from `1` to `0` stops the calls to `Mix()` and allows development of spatial correlations. As seen in the following figures, the overall results change drastically. Figure 5.2 shows the temporal dynamics between the two competing species for the case `dist1=dist2=1`. Whereas in the mixed system the takeover by the dominant species was complete after roughly 100 time units, the takeover is still occurring even after 5000 time units in the spatial case.

These situations call for imaging of the spatial dynamics, if for no other

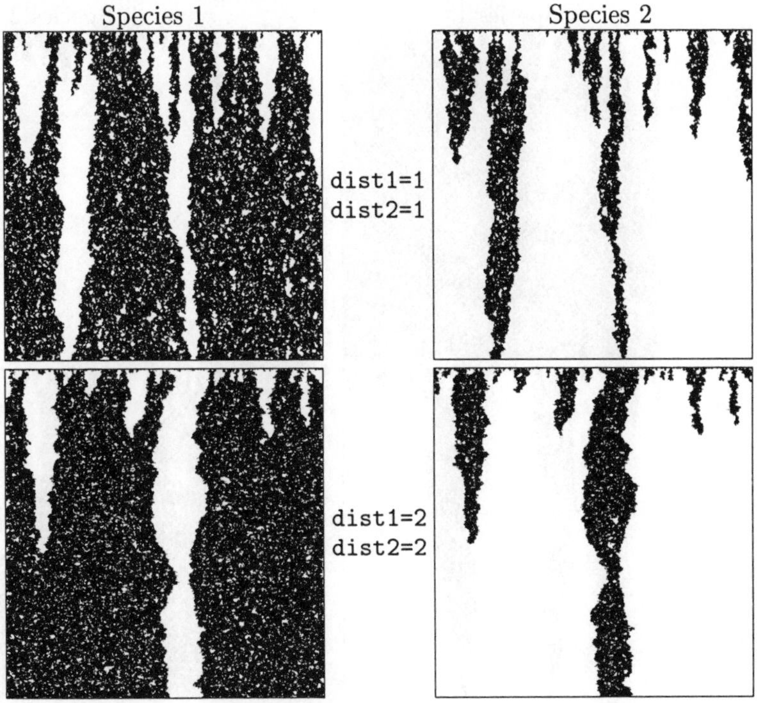

Figure 5.4. Equal offspring dispersal distances for the two competitors. Species 1 wins over species 2 in both cases, but the takeover is drastically slowed from that of the nonspatial system (figure 5.2) due to spatial segregation of the two species. Replacement of one species by the other can only take place at the interface of the two populations.

reason than to see possible reasons for such a drastic slowdown from the nonspatial case. Figure 5.4 plots the spatial dynamics for the two runs, **dist1=dist2=1** and **dist1=dist2=2**. Overall results are the same for the two runs (species 1 wins for this initial condition) but the time required for the takeover is shortened slightly for larger offspring placement distances. In fact, there are two competitive processes taking place here. The first happens very quickly; a localized competition between the two species setting up regions where one species or the other dominates due to local variation in the initial distribution and interactions of the two species. The second competitive process occurs at the distributional boundaries of the two species – we'll examine this process in more detail below.

Figure 5.5 examines the images for the cases of different offspring placement distances. When the dominant species has the longer offspring placement distance, nothing changes except a drastic decrease in the takeover time. The results really change when the poorer competitor species has a

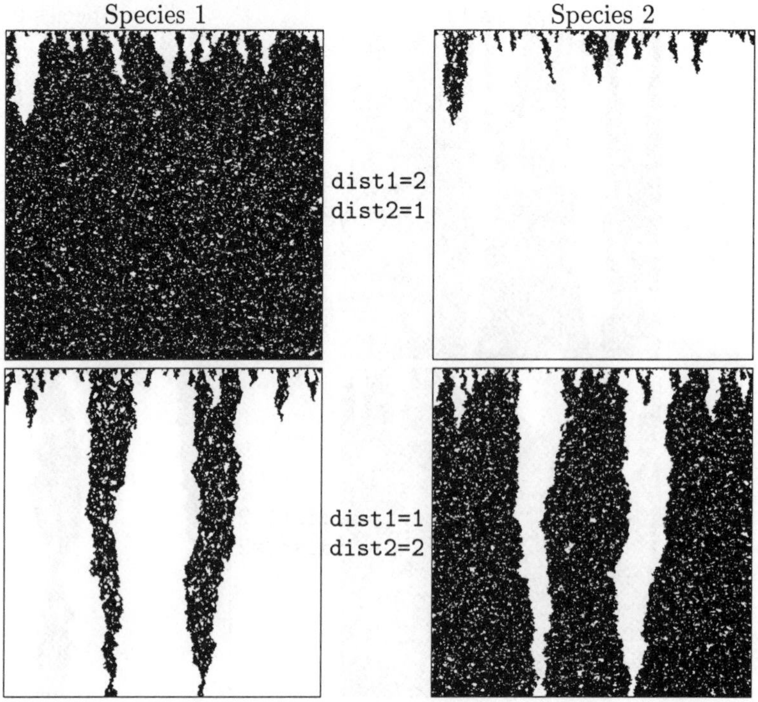

Figure 5.5. Unequal offspring dispersal distances for the two competitors. In the top images, species 1 quickly displaces species 2. Dispersal distances are switched for the run producing the bottom images, and the outcome of competition is completely switched: Species 2 displaces species 1.

larger offspring placement distance – it wins! Poorer competitive ability ($\beta_{12} < \beta_{21}$) can be offset by better dispersal ability to the extent of being a better overall competitor at the distributional interface.

5.3 Analytic Model for the Spatial System

5.3.1 Model Equations

Approaching the spatial system analytically is not a trivial task. Much work has been done on very similar competition problems under various limiting conditions.[2] One mathematical representation for the spatial system follows. Consider a system of L cells arranged in a one-dimensional way with cells at each end connected together. Define state variables for the sites, N^i, that take on values $q = 0, 1, 2$, referring to being empty, or occupied by species 1 or 2, respectively. In light of the two-species competition model simulated

[2] Including Silvertown *et al.* (1992), Molofsky (1994), Wilson and Nisbet (1997), Durrett and Levin (1998), Neuhauser (1998), Holmes and Wilson (1998), and Bolker and Pacala (1999).

above, define additional state variables at each site, N_q^i, representing the probability for site variable N^i to be in state q. At each site, these definitions imply the constraint $N_0^i + N_1^i + N_2^i = 1$, meaning that the site has to be in one of those three states.

We can now represent the dynamics of the site state probabilities as a system of $2L$ equations. The dynamics of the state variables are written as (e.g., Hastings 1980)

$$\begin{aligned} \frac{dN_1^i}{dt} &= \frac{\alpha_1}{2D}(1 - N_1^i - N_2^i) \sum_{j=-D_1}^{j=D_1} N_1^{i+j} \\ &\quad - \frac{1}{2}\beta_{12}(N_1^i N_2^{i-1} + N_1^i N_2^{i+1}) - \mu_1 N_1^i \qquad (5.10a) \\ \frac{dN_2^i}{dt} &= \frac{\alpha_2}{2D}(1 - N_1^i - N_2^i) \sum_{j=-D_1}^{j=D_1} N_2^{i+j} \\ &\quad - \frac{1}{2}\beta_{21}(N_2^i N_1^{i-1} + N_2^i N_1^{i+1}) - \mu_2 N_2^i. \qquad (5.10b) \end{aligned}$$

There are three parts to each equation. As an example consider the first equation: Taking the last term first, when our focal site i is occupied by a 1, it has a rate μ_1 of changing its state to 0. In the middle terms each neighbor in state 2 induces a mortality of $\beta_{12}/2$ on their type 1 neighbors. Here I use the factor of one-half to connect with the nonspatial model (see exercise 5.9). Finally, when the focal site is in state 0, all of its neighbors in state 1 and within range D_1 can throw offspring into it. Also, as in the simulation, I have assumed $K = 1$ individuals per site.

Another mathematical representation for the competition model is a differential equation approach. Essentially, the site indices of (5.10) are replaced by a continuous position variable, x, the state variables become functions of space and time, i.e., $N_1(x,t)$, and the summations over neighboring sites in the reproduction term are replaced by integrals. These replacements give

$$\begin{aligned} \frac{\partial N_1}{\partial t} &= \frac{\alpha_1}{2D}(1 - N_1 - N_2) \int_{x-D_1}^{x+D_1} N_1(y,t)\, dy \\ &\quad - \beta_{12} N_1 N_2 - \mu_1 N_1 \qquad (5.11a) \\ \frac{\partial N_2}{\partial t} &= \frac{\alpha_2}{2D}(1 - N_1 - N_2) \int_{x-D_2}^{x+D_2} N_2(y,t)\, dy \\ &\quad - \beta_{21} N_1 N_2 - \mu_2 N_2. \qquad (5.11b) \end{aligned}$$

There is at least one subtle difference between (5.10) and (5.11): The interference competition terms ignore the nearest-neighbor aspect used in the simulation and described in equation (5.10) (see exercise 5.9). Analytic approaches to these so-called integrodifferential equations (having both derivatives and integrals) are limited because they are just too complicated. (The same holds for the equations governing site–state dynamics.) In chapter 9, however, we examine related equations using approximate methods – specifically, spatial linear stability analysis.

5.3.2 Code Alterations for a Numerical Solution

Here I outline the parts of a C program that provide a numerical solution to a discrete-time representation of (5.11). A similar approach was used by Comins, Hassell, and May (1992) for a deterministic host-parasitoid model, where the goal was not so much to correctly analyze a set of spatially explicit differential equations, but to peruse model behavior. As such, you must be cautious in making conclusions because the approach neither gives correct solutions to continuous-time and space models (Neubert, Kot, and Lewis 1995) nor incorporates stochastic effects (Ruxton and Rohani 1996), both of which can change model predictions.

For the most part, this program is identical to the competition simulation code. I only show the main changes. First, I define several global arrays to hold, respectively, the population density, the locally produced offspring, and the offspring dispersed to a cell, for each species

```
double    n1[CELLS], n2[CELLS];
double    babes1[CELLS], babes2[CELLS];
double    disp1[CELLS],  disp2[CELLS];
```

The heart of the `main()` routine operates on these arrays. I replace the three calls to `Mortality()`, `Competition()`, and `Reproduction()`, in the previous code with the following function calls

```
MakeBabies();
DisperseBabies(disp1,babes1,n1,n2,dist1);
DisperseBabies(disp2,babes2,n1,n2,dist2);
CompeteAndDie();
AddBabies();
```

The function `MakeBabies()` calculates the number of offspring produced at each cell and stores the numbers in the `babes` arrays. `DisperseBabies()` then disperses these offspring to their new cells. `CompeteAndDie()` reduces each cell's population densities according to the competition and mortality terms. Finally, offspring are added to the cells by a call to `AddBabies()`. I list the functions themselves, but as there are no new programming concepts discussion of them is unnecessary.

```
void MakeBabies(void)
{
     int j;
     for(j=0;j<CELLS;j++)
     {
          babes1[j] = alpha1*n1[j];
          babes2[j] = alpha2*n2[j];
     }
}


void CompeteAndDie(void)
{
     int j;
     double oldpop1, oldpop2;

     for(j=0;j<CELLS;j++)
     {
          oldpop1 = n1[j]; oldpop2 = n2[j];
          n1[j] -= (beta12*oldpop2 + mu1)*oldpop1;
          n2[j] -= (beta21*oldpop1 + mu2)*oldpop2;
          if(n1[j]<0.0) n1[j] = 0.0;
          if(n2[j]<0.0) n2[j] = 0.0;
     }
}

void DisperseBabies(double *disp, double *babes,
          double *pop1, double *pop2, int dist)
{
     int j, kmin, k, kmax;
     double babes_per_cell;

     for(j=0;j<CELLS;j++) disp[j] = 0.0;
     for(j=0;j<CELLS;j++)
     {
          babes_per_cell = babes[j]/(2.0*dist+1.0);
          kmin = j - dist;
          if(kmin<0) /* add babies that fall off edges... */
          {         /* ...to the outermost cells */
               disp[0] += (-kmin)*babes_per_cell
                         *(1.0-pop1[0]-pop2[0]);
               kmin = 0;
          }
          kmax = j + dist + 1;
          if(kmax>CELLS)
          {
```

```
                disp[CELLS-1] += (kmax-CELLS)*babes_per_cell
                    *(1.0-pop1[CELLS-1]-pop2[CELLS-1]);
                kmax = CELLS;
            }
            for(k=kmin;k<kmax;k++)
                disp[k] += babes_per_cell*(1.0-pop1[k]-pop2[k]);
        }
}

void AddBabies(void)
{
        int j;
        for(j=0;j<CELLS;j++)
        {
            n1[j] += disp1[j]; n2[j] += disp2[j];
        }
}
```

5.3.3 Numerical Results

Using this crude numerical analysis of the spatial equations, we will examine the dynamics that take place at an interface between regions dominated by the two species. Figure 5.6 presents PostScript images of the results. As in the individual-based simulation, increasing the dispersal distance of species 1 enhances its dominance over species 2. Likewise, increasing the dispersal distance of the "inferior" competitor can overturn these results and allow species 2 to dominate over species 1; however, the effect longer dispersal has on dominance is weaker than in the simulation. We thus have a degree of congruence between the two model formulations.

Clearly, local competition for sites is not the only important biological feature – long-range dispersal ability also plays a role. This conclusion, however, must be examined under the realization that spatial subdivision of resources and concomitant consumer dispersal have long been recognized as important factors in competition models (Shorrocks, Atkinson, and Charlesworth 1979, Hastings 1980, Atkinson and Shorrocks 1981, Ives 1988, Hanski and Zhang 1993, Holmes *et al.* 1994, Pacala and Levin 1997) to the extent that, theoretically, an infinite number of competitors can coexist on as few as two patches (Kishimoto 1990).

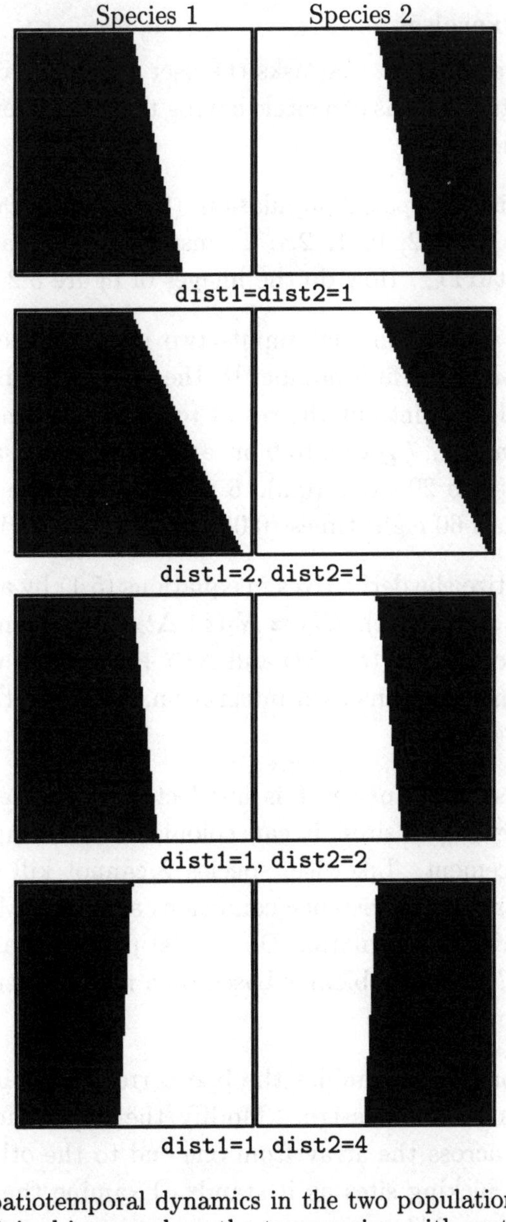

Figure 5.6. Spatiotemporal dynamics in the two populations for the analytic model. Paired images show the two species, with systems initialized having species 1 on the left and species 2 on the right. If a species' density in a cell exceeds 0.5, then the corresponding pixel in its image is dark.

5.4 Exercises

5.1 Write a program that asks the user to input a `double` number, then outputs the areas of a circle having that diameter and a square having that side.

5.2 Initialize the spatial populations (the `state` array) as the nonrandom series, 0, 1, 2, 0, 1, 2, ..., instead of the random series used in `InitStuff()`. How do the images of figure 5.2 change, if at all?

5.3 Write a program that inputs two integers, then performs the long division of the first number by the second number using only integer variables. Print out the result to twenty decimal places. For example, 9 and 7. 7 goes into 9 once (1). 2 remains; add the tenths place. 7 goes into 20 twice (0.2). 6 remains; add the hundredths place. 7 goes into 60 eight times (0.08). And so on, giving 1.2857...

5.4 Discretize the derivatives in equations (5.1) by approximating $dN_1 \approx N_1(t+\Delta t) - N_1(t)$, $dN_2 \approx N_2(t+\Delta t) - N_2(t)$, and $dt \approx \Delta t$. Solve the equations for $N_1(t+\Delta t)$ and $N_2(t+\Delta t)$, then write a program that performs the numerical integration. Examine the results for several values of Δt.

5.5 Suppose that species 1 is unaffected by species 2. As well as colonizing empty sites, it can colonize sites occupied by species 2 by displacement. Likewise, species 2 cannot kill species 1 ($\beta_{12} = 0$). How are the coexistence conditions affected? Modify the simulation to model this scenario. Do spatial processes alter coexistence outcomes? (This problem is based on a model discussed by Holmes and Wilson (1998).)

5.6 This problem examines the bad correlations that can be set up by poor algorithm design. Modify the `Reproduction()` function to sweep across the array from one end to the other, rather than randomly picking sites as it stands. Examine the PostScript files – is anything odd?

5.7 Initialize the lower half of the lattice with members of species 1 and the upper half with members of species 2. This initialization creates a single interface between the two populations. Write a measurement algorithm that locates the position of the interface. The slope of the interface position versus time curve gives the speed of the interface.

Examine the interface speed as a function of the important simulation parameters (see Ellner *et al.* 1998).

5.8 Modify the simulation code as described in section 5.3.2 to produce the images shown in figure 5.6.

5.9 The function `CompeteAndDie()` on page 121 used within-cell populations to calculate interference competition induced mortality, rather than using neighboring cell populations as done in the individual-based simulation and described in equation (5.10). Make the appropriate modifications to `CompeteAndDie()` to account for this detail; are the numerical results shown in figure 5.6 affected?

6

Programming Projects

Programming components covered in this chapter:
• top–down program writing • debugging hints • `switch` statement

In this chapter I demonstrate writing two simulations from scratch, each starting with a written project description. One good scientific programming approach begins by specifying the broad concepts and progressively working down to detailed functions. An analogy to this process is the idealized approach to writing a manuscript – you start with an outline, perhaps some figures and equations, then start wrapping words around the major points. A top–down programming style is similar in that you flesh out the most important aspects first, then incorporate the details later. This approach forces you to think carefully about higher-level concepts before being buried in details. This approach will change as you become a more proficient programmer: Having built up a large stock of code living on your hard-drive, you'll pull code from old projects to use as the foundation for new projects. A new project might involve only changing detailed functions.

Coding up a problem is a highly variable, personal task. It would be highly unlikely that two people, given the same problem description, independently produced identical programs. This variability is acceptable because many different computer programs can lead to the same results – the problem arises when different computer programs, presumably having identical assumptions, produce different results. Hence, benchmarking simulation results against some "independent," perhaps analytical, results is a dominant theme in this book.

6.1 Metapopulation Dynamics

6.1.1 Problem Statement

What are the mutual and competing effects of local reproduction and dispersal within a metapopulation? Imagine a one-dimensional line of ephemeral resource patches. Imagine that insects live on these patches, but the insects live for several time steps. During a time step the population at a patch experiences discrete-time logistic growth: $N_{t+\Delta t} = N_t + r\Delta t N_t(1 - N_t/K)$. If the population in a patch becomes negative, set it to zero. At any particular time step, every other patch can be occupied, and each patch is occupiable for exactly one time step. Thus, during one time the patch configuration is `XOXOXOXOXO...` with `O` being occupiable, and the next time step the patch configuration is `OXOXOXOXOX...`, flipping back to the first configuration the next time step. Use boundary conditions that make patches at the ends be neighbors. When each patch dies, a random number, $0 \leq q < 1$, determines the population fraction moving to the new patch on the left and the rest moving to the right. Each patch has a new q each time step. Fix $K = 10$ and start with each occupiable patch having K insects. (This model has its conceptual origins in the *q-model* for sandpiles (Coppersmith *et al.* 1995).)

6.1.2 Coding the Project

Here I demonstrate top–down programming: While coding up the problem, initially we won't worry about details such as function prototyping and measuring system properties – that will all fall out as we refine the program and figure out the important details. The priority at this stage is getting the simulation written. Discrete-time updating is one of the most obvious features of the project, implying that the main loop is over some maximum number of time steps `MAXSTEPS`. Thus we start with

```
main()
{
     for(t=0;t<MAXSTEPS;t++)
     {
     }
}
```

I left `t` undeclared and `MAXSTEPS` undefined, and will fill in those details later. Now list the processes that occur within a time step

```
main()
{
     for(t=0;t<MAXSTEPS;t++)
```

```
    {
        MoveInsects();
        ReproduceInsects();
    }
}
```

Ordering of these processes seems irrelevant at this point,[1] but let's follow the project description. First think about `MoveInsects()`

```
MoveInsects()
{
    for(ipat=0;ipat<NPATCHES;ipat++)
    {
    }
}
```

because each of the `NPATCHES` patches must be dealt with individually. At each patch the insects fly up, segregate into two groups, then land on the two neighboring patches

```
MoveInsects()
{
    for(ipat=0;ipat<NPATCHES;ipat++)
    {
        qq = drand48();
        leftpatch += oldpatch*qq;
        rightpatch += oldpatch*(1.0-qq);
    }
}
```

where `oldpatch` refers to the old patch's population, and `leftpatch` and `rightpatch` refer to the neighboring patches' populations. Why the `+=`? Each new patch receives insects from two old patches and it remains unclear which of the two is dealt with first. This updating also points out that initialization to zero must be performed.

One big problem here is that I used `NPATCHES` as the number of occupied patches, and as the problem description states, only every other patch is occupied during any one time step. This one-half occupancy means that the total number of patches in the system, what I had meant as `NPATCHES`, must be an even number. So, more carefully defining the patches and rewriting the array references

```
MoveInsects()
{
    double patch[NPATCHES];
```

[1] The ordering of interactions can have important implications – see McCauley, Wilson, and de Roos (1993).

```
    for(ipat=0;ipat<NPATCHES;ipat+=2)
    {
        qq = drand48();
        patch[ipat-1] += patch[ipat]*qq;
        patch[ipat+1] += patch[ipat]*(1.0-qq);
    }
}
```

That's better, but there are still two big problems. The first problem deals with the wrapped (also called periodic) boundary conditions. Boundary problems are clear when `ipat` is zero – we try to access `patch[-1]`, an action the compiler probably accepts, but when running the program will cause big problems.[2] The second problem is apparent when looking ahead to the next time step. Moving the population then requires that we start with the population on patch 1, not patch 0. Resolving this latter problem demands that the calling function tell `MoveInsects()` which sublattice to work on, meaning it needs to be passed an argument

```
MoveInsects(int ilat)
{
    double patch[NPATCHES];

    for(ipat=ilat;ipat<NPATCHES;ipat+=2)
    {
        qq = drand48();
        patch[ipat-1] += patch[ipat]*qq;
        patch[ipat+1] += patch[ipat]*(1.0-qq);
    }
}
```

where `ilat` has the value `0` or `1` depending on which of the two sublattices `MoveInsects()` is supposed to deal with. Now let's deal with the boundary problems. If `ilat=0`, then there is a problem with the first patch because we want to reference the last patch in the array as the new patch on the left. Alternatively, if `ilat=1`, then when we have a problem with the last patch (when `ipat=NPATCHES-1`) because the new patch on the right should be the first patch of the array, `patch[0]`. These two patches are special cases

```
MoveInsects(int ilat)
{
    double patch[NPATCHES];

    if(ilat==0)
    {
```

[2] C compilers are notorious for not catching programming errors, or, if they do, not giving very helpful clues as to the nature of the problem. I demonstrate a crude debugging approach as we go.

```
        qq = drand48();
        patch[NPATCHES-1] += patch[0]*qq;
        patch[1]          += patch[0]*(1.0-qq);
    }
    for(ipat=2-ilat;ipat<NPATCHES-1;ipat+=2)
    {
        qq = drand48();
        patch[ipat-1] += patch[ipat]*qq;
        patch[ipat+1] += patch[ipat]*(1.0-qq);
    }
    if(ilat==1)
    {
        qq = drand48();
        patch[NPATCHES-2] += patch[NPATCHES-1]*qq;
        patch[0]          += patch[NPATCHES-1]*(1.0-qq);
    }
}
```

At this point I think the problems outlined above are resolved. The first patch and last patch are dealt with and I modified the loop over the other patches accounting for their special treatment.

Now let's make several needed changes: 1) define local variables, 2) recognize that we reference the **patch[]** array throughout the program and make it a globally declared array, and 3) initialize the new patches within **MoveInsects()**. These modifications give for the entire program so far

```
double patch[NPATCHES];

main()
{
    for(t=0;t<MAXSTEPS;t++)
    {
        MoveInsects(t%2);
        ReproduceInsects();
    }
}

MoveInsects(int ilat)
{
    int ipat;
    double qq;

    for(ipat=1-ilat;ipat<NPATCHES;ipat+=2)
        patch[ipat] = 0.0;

    if(ilat==0)
    {
        qq = drand48();
        patch[NPATCHES-1] += patch[0]*qq;
```

```
        patch[1]          += patch[0]*(1.0-qq);
    }
    for(ipat=2-ilat;ipat<NPATCHES-1;ipat+=2)
    {
        qq = drand48();
        patch[ipat-1] += patch[ipat]*qq;
        patch[ipat+1] += patch[ipat]*(1.0-qq);
    }
    if(ilat==1)
    {
        qq = drand48();
        patch[NPATCHES-2] += patch[NPATCHES-1]*qq;
        patch[0]          += patch[NPATCHES-1]*(1.0-qq);
    }
}
```

Now turn our attention to ReproduceInsects(). This function updates local patch populations via a deterministic population change equation contained within some other as-of-yet-to-be-defined function NewPop()

```
ReproduceInsects(int ilat)
{
    for(ipat=ilat;ipat<NPATCHES;ipat+=2)
    {
        patch[ipat] = NewPop(patch[ipat]);
    }
}
```

In this function, ilat references the new sublattice (occupied after the insects moved) rather than the old sublattice (occupied before the insects moved). Within one time step I prefer referencing a single sublattice rather than multiple sublattices, thus, I modify the main() routine

```
main()
{
    for(t=0;t<MAXSTEPS;t++)
    {
        ReproduceInsects(t%2);
        MoveInsects(t%2);
    }
}
```

Back to work on ReproduceInsects(). It calls NewPop(), which updates a population according to the equation given in the problem description

```
double NewPop(double pop)
{
    double newpop;

    newpop = pop + rr*pop*(1.0-pop/kk);
```

```
    if(newpop<0.0) newpop = 0.0;
    return(newpop);
}
```

We're almost done. The two functions called in `main()` are finished, except for the declaration of a few global variables, and initializations of variables and arrays must be done. The standard libraries can be included. We must also initialize the random number generator. Adding these details gives

```
#include <stdlib.h>
#include <stdio.h>
#include <math.h>

#define NPATCHES    20
#define MAXSTEPS    50
#define RR          0.1
#define KK          10.0
#define DELT        1.0

double rr, kk;
double patch[NPATCHES];
void ReproduceInsects(int);
void MoveInsects(int);

int main(void)
{
    int t, ipat, seed;

    seed = 1456739853;
    srand48(seed);

    rr = RR*DELT;
    kk = KK;
    for(ipat=0;ipat<NPATCHES;ipat++)
        patch[ipat] = kk;

    for(t=0;t<MAXSTEPS;t++)
    {
        ReproduceInsects(t%2);
        MoveInsects(t%2);
    }
    return (0);
}
```

There is only one problem now – what do we measure? At first, just to get a feel for what goes on in the system, we might find something interesting by printing the first few patch populations on even time steps. To do so add a new function, `OutputInsects()`

```
void OutputInsects(int time, int ilat)
{
     int ipat;

     if(ilat==0)
          for(ipat=0;ipat<16;ipat+=2)
               printf("%f",patch[ipat]);
}
```

I started writing this function by copying the `ReproduceInsects()` function, then editing it. I noticed that `ipat` was undeclared, so add that declaration to `ReproduceInsects()`. We also want the time output (which must be passed as an argument to `OutputInsects()`) and a return at the end of each line.

6.1.3 The Finished Product

Well, finished might not be the most appropriate term, but we have a complete program.

```
#include <stdlib.h>
#include <stdio.h>
#include <math.h>

#define NPATCHES     20
#define MAXSTEPS     50
#define RR           0.1
#define KK           10.0
#define DELT         1.0

double rr, kk;
double patch[NPATCHES];
void OutputInsects(int, int);
void ReproduceInsects(int);
void MoveInsects(int);

int main(void)
{
     int t, ipat, seed;

     seed = 1456739853;
     srand48(seed);

     rr = RR*DELT;
     kk = KK;
     for(ipat=0;ipat<NPATCHES;ipat++)
          patch[ipat] = kk;
```

```
    for(t=0;t<MAXSTEPS;t++)
    {
        OutputInsects(t,t%2);
        ReproduceInsects(t%2);
        MoveInsects(t%2);
    }
    return (0);
}

void OutputInsects(int time, int ilat)
{
    int ipat;

    if(ilat==0)
    {
        printf("%4d ",time);
        for(ipat=0;ipat<16;ipat+=2)
            printf("%8.4f ",patch[ipat]);
        printf("\n");
    }
}

void ReproduceInsects(int ilat)
{
    int ipat;

    for(ipat=ilat;ipat<NPATCHES;ipat+=2)
    {
        patch[ipat] = NewPop(patch[ipat]);
    }
}

double NewPop(double pop)
{
    double newpop;

    newpop = pop + rr*pop*(1.0-pop/kk);
    if(newpop<0.0) newpop = 0.0;
    return(newpop);
}

void MoveInsects(int ilat)
{
    int ipat;
    double qq;

    for(ipat=1-ilat;ipat<NPATCHES;ipat+=2)
        patch[ipat] = 0.0;

    if(ilat==0)
```

```
    {
        qq = drand48();
        patch[NPATCHES-1] += patch[0]*qq;
        patch[1]          += patch[0]*(1.0-qq);
    }
    for(ipat=2-ilat;ipat<NPATCHES-1;ipat+=2)
    {
        qq = drand48();
        patch[ipat-1] += patch[ipat]*qq;
        patch[ipat+1] += patch[ipat]*(1.0-qq);
    }
    if(ilat==1)
    {
        qq = drand48();
        patch[NPATCHES-2] += patch[NPATCHES-1]*qq;
        patch[0]          += patch[NPATCHES-1]*(1.0-qq);
    }
}
```

6.1.4 Compiling, Debugging, and Running

On a Unix machine I compiled the program with the line

```
cc metapop.c -o bmp -lm
```

and the computer returned

```
"metapop.c", line 64: identifier redeclared: NewPop
     current : function(double) returning double
     previous: function() returning int : "metapop.c", line 59
cc: acomp failed for metapop.c
```

Ahhh, the joys of debugging! Looking at the `NewPop()` function everything looks OK, but I forgot to declare it as type `double` up at the top of the program prior to `main()`. Doing so and recompiling yields no complaints from the compiler. Of course, success often gives a false sense of security!

Running the program gives the following output

```
 0  10.0000  10.0000  10.0000  10.0000  10.0000  10.0000  10.0000  10.0000
 2  15.3658  12.7048   6.2268  10.9534  14.7090   5.2655   7.0288  20.5733
 4   9.5724   9.8461   8.1996   7.2073  19.1481   8.4432   1.9135  14.7946
 6   6.8207   3.9707   4.9917  15.2380  11.1925  10.4341   4.9386   2.8522
 8   9.0835   2.5403   7.3082  15.3859  15.7269   2.0837   6.3256   0.4285
10  16.3856   1.0620  11.8747  17.5893   7.6176   3.6375   2.9962   4.2446
12   2.4377  16.0090   6.0690  24.4487   3.7096   4.9024   2.1637   7.5062
14   4.0391  11.4169   9.7203  11.7591  13.3670   3.4188   0.9047   8.4320
16   5.3364   7.6065   7.0140  25.3091   7.1629   1.9405   1.7568   2.3237
18   9.7504   5.7080  19.8772   2.7887  12.3703   4.3500   4.4185   5.9614
20  15.9857   7.7764  11.4833   7.8570   7.0522   5.8945   6.6726   3.7102
22  10.3766  15.3019  11.1246   6.3090   4.7914   1.8971   8.9303   5.5994
```

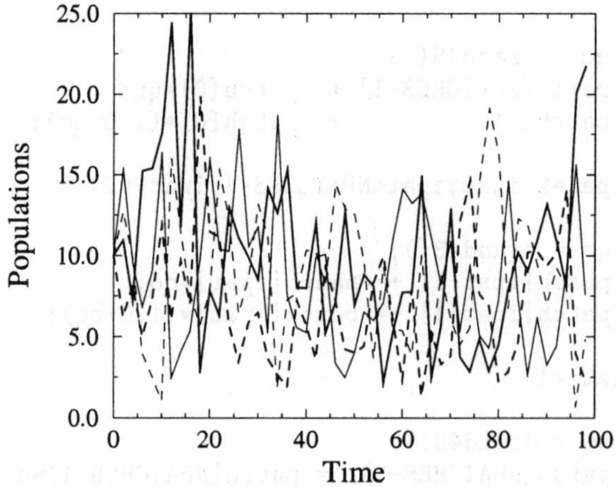

Figure 6.1. Results of the metapopulation model. Population numbers for the first four occupied patches on even time steps are plotted as a function of time. Dynamics of individual patches, even close ones, are extremely variable.

24	10.2738	12.1309	5.7353	12.9535	5.3708	4.1198	13.3371	0.7091
26	17.8695	6.4344	3.4690	11.0918	8.0979	5.7775	7.9076	7.8312
28	10.5625	8.6435	5.8923	9.8988	5.2988	11.1715	12.5602	5.0978
30	11.7887	13.1561	6.4903	8.5293	4.5557	10.1429	10.4318	3.1834
32	5.1994	11.1928	3.6762	14.1539	9.9051	3.4558	5.1384	8.8008
34	18.7260	1.8365	2.7091	12.6143	6.8815	7.8826	7.4951	6.4826
36	7.5500	7.3051	1.9005	15.3363	4.5724	3.3693	7.2195	11.2861
38	5.5431	7.6736	7.2493	8.0273	3.3889	8.8673	7.5286	13.7613
40	5.2624	10.3492	5.7229	7.9827	6.8418	6.0831	8.9043	14.6042
42	9.9263	10.2062	3.6139	12.0945	3.6798	7.1890	11.7069	13.9918
44	10.0833	8.2169	9.2911	5.2189	5.0668	9.8130	17.4196	6.2716
46	3.3767	14.4468	7.2167	7.1122	6.2210	7.7338	9.5171	14.6598
48	2.4832	12.8901	4.2419	12.7408	12.0188	2.5893	6.8283	13.6441

This quick and dirty output is a good way to see at a glance if a program is at least functioning, but an inadequate way to visualize results. The numbers are plotted in figure 6.1.

Remember that the population dynamics at an individual patch is the discrete-time logistic growth model with a low growth parameter, which means that the populations would relax to the carrying capacity K in the absence of dispersal. Random dispersal between patches pushes local patch populations above and below the carrying capacity, keeping all patches perpetually away from K. These mutual and competing effects of reproduction and dispersal may be interesting in the context of population growth on ephemeral patches. The hard work – asking the right ecological questions, measuring the right quantities, and putting it all into a broader theoretical framework – is left as an exercise for the reader.

6.2 Disease Dynamics

6.2.1 Problem Statement

How are disease dynamics affected by localized transmission in a spatial system? Imagine a one-dimensional line of infinitely lived organisms (or quickly replaced shorter-lived ones) with periodic boundary conditions. A disease affects this species, such that each organism is in one of four basic states: (S) susceptible to the disease; (E) exposed to the disease but not yet infectious to other organisms; (I) infectious to other organisms; and (R) recovered and immune to the disease. State S lasts until an organism is exposed to the disease, but each of the other states, E, I, and R, last for times T_{E}, T_{I}, T_{R}. Once the time T_{R} is up for an organism in state R it goes back to the susceptible state S and can once again catch the disease.

6.2.2 Coding the Project

Again I demonstrate the top–down programming approach. This model appears time-driven, thus the outer loop will run over time steps

```
main()
{
     for(i=0;i<MAXTIME;i++)
     {
     }
}
```

At this point it seems that we need two basic functions accounting for the most important processes, spreading the disease between organisms and changing organisms' states

```
main()
{
     for(i=0;i<MAXTIME;i++)
     {
          SpreadDisease();
          ChangeStates();
     }
}
```

Let's first consider the `SpreadDisease()` function. We need to assume something about the transmission distance for the disease: For example, to how many cells away can an infected organism pass on its infection? In this example I assume that an organism can only catch the disease from its immediate neighbors

```
SpreadDisease()
{
     if(state[0]==HEALTHY) StoE(0,CELLS-1,1);
     for(i=1;i<CELLS-1;i++)
          if(state[i]==HEALTHY) StoE(i,i-1,i+1);
     if(state[CELLS-1]==HEALTHY) StoE(CELLS-1,CELLS-2,0);
}

StoE(int cell, int cellL, int cellR)
{
     int cnt=0;

     if(state[cellL]==INFECTED) cnt = 1;
     if(state[cellR]==INFECTED) cnt++;
     if(cnt)
     {
          if(drand48()<(double)cnt*inf_rate)
          {
               state[cell] = EXPOSED;
               timer[cell] = EXP_TIME;
          }
     }
}
```

with the following defined constants and variables

```
#define   HEALTHY      0
#define   EXPOSED      1
#define   INFECTED  2
#define   IMMUNE       3

#define   EXP_TIME  3
#define   INF_TIME  3
#define   IMM_TIME  3

int  state[CELLS], timer[CELLS];
```

The first four constants are proxies for the state of an organism, enabling greater legibility of the program. For example, there is no difference between `if(state[cellL]==INFECTED)` and `if(state[cellL]==2)` from the computer's point of view, but from the perspective of a human reading the code the first form is much clearer. Array `state[i]` holds the disease state of organism `i`. The `_TIME` constants define the exposure, infection, and immune times, and `timer[i]` keeps track of this time for organism `i`.

The next task is writing function `ChangeStates()`. The processes addressed here are purely within-cell interactions and do not involve cell-to-cell dynamics:

```
ChangeStates()
{
     for(i=0;i<CELLS;i++)
     {
          switch(state[i])
          {
               case IMMUNE:
                    timer[i]--;
                    if(timer[i]==0)
                         state[i] = HEALTHY;
                    break;

               case INFECTED:
                    timer[i]--;
                    if(timer[i]==0)
                    {
                         state[i] = IMMUNE;
                         timer[i] = IMM_TIME;
                    }
                    break;

               case EXPOSED:
                    timer[i]--;
                    if(timer[i]==0)
                    {
                         state[i] = INFECTED;
                         timer[i] = INF_TIME;
                    }
                    break;

               default:
          }
     }
}
```

The `switch` statement is a simpler, cleaner way of writing a long series of `if-else` statements. This line acts as a switch, passing control to the `case` statement matching the argument of the `switch` statement. Thus, when `state[i]` equals `INFECTED`, control passes to the `case INFECTED` statement. The computer then executes the block of code after the colon. The `break` statement drops us out of the `switch` statement, falling back to the `for` loop and on to processing the next cell. Without the `break` statement, execution proceeds with the lines in the next `case` statement. If `state[i]` was not one of the cases shown (i.e., `HEALTHY`), control passes to the `default` case (here it does nothing).

The program is nearly finished except for some details like local variable definitions and initialization of arrays. Initialization occurs in a function

`InitStuff()` called from `main()`. A reasonable initial condition is a susceptible population containing a few exposed organisms.

```
InitStuff()
{
	unsigned long seed=145739853;
	srand48(seed);

	for(i=0;i<CELLS;i++)
	{
		state[i] = HEALTHY;
		timer[i] = 0;
		if(drand48()<INIT_EXP)
		{
			state[i] = EXPOSED;
			timer[i] = EXP_TIME * drand48() + 1;
		}
	}
}
```

Last, but not least, something should be measured. An initial examination might measure the frequencies of each disease state

```
Measurements()
{
	scnt = ecnt = icnt = rcnt = 0.0;

	for(i=0;i<CELLS;i++)
	{
		if(state[i]==HEALTHY)           scnt += 1.0;
		else if(state[i]==EXPOSED)      ecnt += 1.0;
		else if(state[i]==INFECTED)     icnt += 1.0;
		else if(state[i]==IMMUNE)       rcnt += 1.0;
	}
	scnt /= (double)(CELLS);
	ecnt /= (double)(CELLS);
	icnt /= (double)(CELLS);
	rcnt /= (double)(CELLS);
}
```

then output those numbers to the terminal.

6.2.3 The Finished Product

Adding all the details of local variable definitions and such leads to the complete program.

```
#define   HEALTHY        0
#define   EXPOSED        1
#define   INFECTED  2
#define   IMMUNE         3

#define   EXP_TIME  3
#define   INF_TIME  3
#define   IMM_TIME  3

#define   INF_RATE  (0.35)
#define   INIT_EXP  (0.1)

#define   MAXTIME        100
#define   CELLS          1000

int       state[CELLS], timer[CELLS];
double    inf_rate, scnt, ecnt, icnt, rcnt;
void InitStuff();
void SpreadDisease(void);
void StoE(int, int, int);
void ChangeStates(void);
void Measurements(void);

int main(void)
{
     int i;

     InitStuff();
     for(i=0;i<MAXTIME;i++)
     {
          SpreadDisease();
          ChangeStates();
          Measurements();
          printf("%d %f %f %f %f\n", i, scnt, ecnt, icnt, rcnt);
     }
     return (0);
}

void InitStuff(void)
{
     int i;
     unsigned long seed=145739853;
     srand48(seed);

     inf_rate = INF_RATE;
```

```
    for(i=0;i<CELLS;i++)
    {
        state[i] = HEALTHY;
        timer[i] = 0;
        if(drand48()<INIT_EXP)
        {
            state[i] = EXPOSED;
            timer[i] = EXP_TIME * drand48() + 1;
        }
    }
}

void SpreadDisease(void)
{
    int i;

    if(state[0]==HEALTHY) StoE(0,CELLS-1,1);
    for(i=1;i<CELLS-1;i++)
        if(state[i]==HEALTHY) StoE(i,i-1,i+1);
    if(state[CELLS-1]==HEALTHY) StoE(CELLS-1,CELLS-2,0);
}

void StoE(int cell, int cellL, int cellR)
{
    int cnt=0;

    if(state[cellL]==INFECTED) cnt = 1;
    if(state[cellR]==INFECTED) cnt++;
    if(cnt)
    {
        if(drand48()<(double)cnt*inf_rate)
        {
            state[cell] = EXPOSED;
            timer[cell] = EXP_TIME;
        }
    }
}

void ChangeStates(void)
{
    int i;

    for(i=0;i<CELLS;i++)
    {
        switch(state[i])
        {
            case IMMUNE:
                timer[i]--;
```

```
                if(timer[i]==0)
                    state[i] = HEALTHY;
                break;

            case INFECTED:
                timer[i]--;
                if(timer[i]==0)
                {
                    state[i] = IMMUNE;
                    timer[i] = IMM_TIME;
                }
                break;

            case EXPOSED:
                timer[i]--;
                if(timer[i]==0)
                {
                    state[i] = INFECTED;
                    timer[i] = INF_TIME;
                }
                break;

            default:
        }
    }
}

void Measurements(void)
{
    int i;

    scnt = ecnt = icnt = rcnt = 0.0;

    for(i=0;i<CELLS;i++)
    {
        if(state[i]==HEALTHY)          scnt++;
        else if(state[i]==EXPOSED)     ecnt++;
        else if(state[i]==INFECTED)    icnt++;
        else if(state[i]==IMMUNE)      rcnt++;
    }
    scnt /= (double)(CELLS);
    ecnt /= (double)(CELLS);
    icnt /= (double)(CELLS);
    rcnt /= (double)(CELLS);
}
```

6.2.4 Compiling, Debugging, and Running

I compiled the program with the line

```
cc seir.c -o bseir -lm
```

and the computer was kind enough not to complain. When I ran the program, the following numbers were output

```
0 0.000000 0.000000 1.000000 0.000000
1 0.000000 0.000000 1.000000 0.000000
2 0.000000 0.000000 1.000000 0.000000
3 0.000000 0.000000 0.000000 1.000000
4 0.000000 0.000000 0.000000 1.000000
5 0.000000 0.000000 0.000000 1.000000
6 1.000000 0.000000 0.000000 0.000000
7 1.000000 0.000000 0.000000 0.000000
8 1.000000 0.000000 0.000000 0.000000
9 1.000000 0.000000 0.000000 0.000000
10 1.000000 0.000000 0.000000 0.000000
```

and so on until time step 100. Clearly there is a problem! The biggest clue is that I intended to initialize only 10 percent of the population as exposed while the rest of the population was susceptible. The numbers show all the organisms as being infected right away. Let's examine that problem by first checking the population counts immediately after `InitStuff()`. Hence, change the `main()` routine to

```
int main(void)
{
      int i;
      InitStuff();
      Measurements();
      printf("%f %f %f %f\n",
            scnt, ecnt, icnt, rcnt);

      for(i=0;i<MAXTIME;i++)
      {
            SpreadDisease();
            ChangeStates();
            Measurements();
            printf("%d %f %f %f %f\n",
                  i, scnt, ecnt, icnt, rcnt);
      }
      return (0);
}
```

Recompiling and running gives the following output

```
0.000000 1.000000 0.000000 0.000000
0 0.000000 0.000000 1.000000 0.000000
```

```
1 0.000000 0.000000 1.000000 0.000000
2 0.000000 0.000000 1.000000 0.000000
3 0.000000 0.000000 0.000000 1.000000
4 0.000000 0.000000 0.000000 1.000000
5 0.000000 0.000000 0.000000 1.000000
6 1.000000 0.000000 0.000000 0.000000
7 1.000000 0.000000 0.000000 0.000000
8 1.000000 0.000000 0.000000 0.000000
9 1.000000 0.000000 0.000000 0.000000
```

The problem is within `InitStuff()` and the way it initializes states. We can check our initializations by changing `InitStuff()`'s inner loop to

```
     for(i=0;i<CELLS;i++)
     {
          state[i] = HEALTHY;
          timer[i] = 0;
          if(drand48()<INIT_EXP)
          {
printf("E");
               state[i] = EXPOSED;
               timer[i] = EXP_TIME * drand48() + 1;
          }
else printf("S");
     }
printf("\n");
```

These `printf()` statements are simple debugging tools tracing program flow and, hopefully, identifying the problem. I usually don't indent these statements to the same level as the other statements – I treat them on a temporary basis and the lack of indentation helps me see them after debugging the code when I want to delete these lines. Admittedly, these debugging statements are crude but still very useful; more sophisticated debugging tools and programs exist allowing examination of variables and such while the program is running, but those debuggers are rather involved. I favor these crude methods and never use debuggers. Recompiling and running gives the following output

```
ESSSSSSSSSSSSSSSSSSSSSSSSSSSSSSSSSSSSSSSSSSSSSSSSSSSSSSSS
SSSSSSSSSSSSSSSSSSSSSSSSSSSSSSSSSSSSSSSSSSSSSSSSSSSSSSSSS
SSSSSSSSSSSSSSSSSSSSSSSSSSSSSSSSSSSSSSSSSSSSSSSSSSSSSSSSS
SSSSSSSSSSSSSSSSSSSSSSSSSSSSSS...<etc.>
0.999000 0.001000 0.000000 0.000000
0 0.999000 0.001000 0.000000 0.000000
1 0.999000 0.001000 0.000000 0.000000
2 0.999000 0.001000 0.000000 0.000000
3 0.999000 0.000000 0.001000 0.000000
4 0.997000 0.002000 0.001000 0.000000
5 0.997000 0.002000 0.001000 0.000000
```

```
6 0.997000 0.000000 0.002000 0.001000
7 0.995000 0.002000 0.002000 0.001000
8 0.995000 0.002000 0.002000 0.001000
9 0.996000 0.000000 0.002000 0.002000
```

Well, that output is odd. Previously everything was set to `EXPOSED`, but now everything but the first state is set to `S`. Adding these lines of code that print out the letters caused a change in the initialization that resulted in the new numbers. This difference indicates that something is going wrong within the test of the `if()` statement. The test compares two `double` quantities, a constant and a random number from `drand48()` – aha! – the problem is with `drand48()`. I forgot to include the usual set of header files. I have made this mistake before, and the first time I did it, it took me a lot longer to find the mistake. The problem can be corrected simply by including the standard library via the line

```
#include <stdlib.h>
```

Making this modification, removing the debugging statements, recompiling, and running gives

```
0 0.918000 0.055000 0.027000 0.000000
1 0.903000 0.043000 0.054000 0.000000
2 0.878000 0.040000 0.082000 0.000000
3 0.840000 0.063000 0.070000 0.027000
4 0.820000 0.058000 0.068000 0.054000
5 0.797000 0.043000 0.078000 0.082000
6 0.808000 0.039000 0.083000 0.070000
7 0.808000 0.043000 0.081000 0.068000
8 0.811000 0.052000 0.059000 0.078000
9 0.807000 0.044000 0.066000 0.083000
10 0.818000 0.033000 0.068000 0.081000
```

and so on. Again, numbers are not a good way to examine output, so the data are plotted in figure 6.2. Apparently the disease dies out. Just to be sure things are working correctly, figure 6.3 presents some PostScript images of the system using modified versions of the Encapsulated PostScript file functions introduced in chapter 4.

Things look OK from a spatial perspective. We can try to make the disease persist by changing parameter values. Figure 6.4A increases the infection rate to 0.45, and in figure 6.4B the infection time is doubled to six time steps. In both cases the disease persists in the host population.[3]

[3] If your interest is piqued by spatial disease models, two very interesting recent references include pair approximation analysis by Satō, Matsuda, and Sasaki (1994) and a reaction–diffusion analysis with empirical connections by Dwyer (1992).

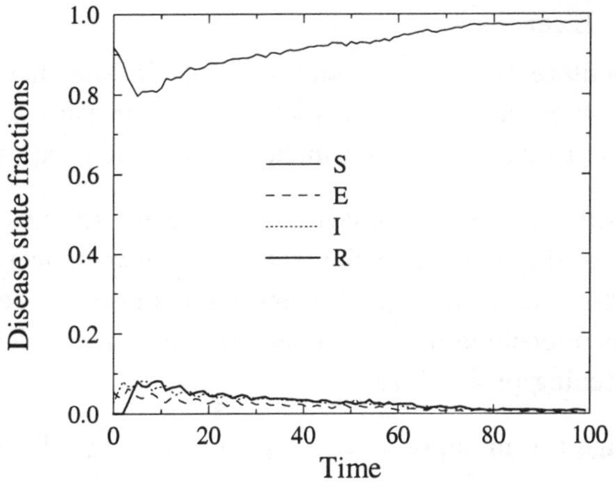

Figure 6.2. State dynamics for the SEIR disease simulation project using the parameter values listed in the code (including INF_RATE=0.35). The disease does not persist.

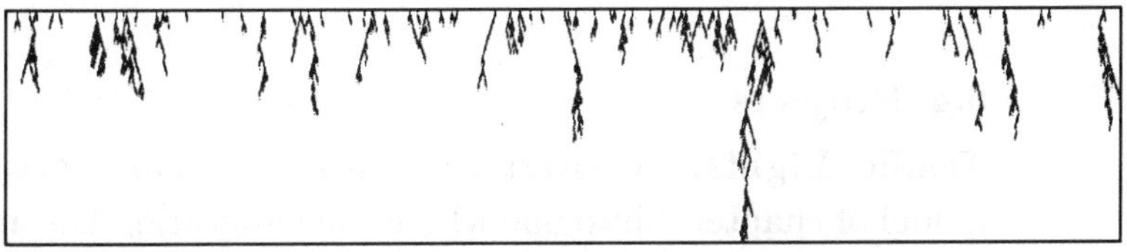

Figure 6.3. PostScript image of disease state dynamics for 100 time steps using the parameter values listed in the code (including INF_RATE=0.35). Infected individuals are dark, exposed grey, and immune light grey. Healthy individuals fill the rest of space and are not shown. The system width is 1000 cells.

(A)

(B)

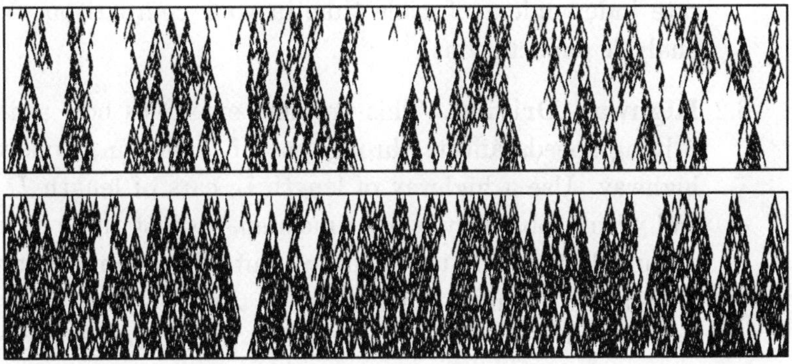

Figure 6.4. PostScript images of disease state dynamics as in figure 6.3 but with (A) an increased infection rate INF_RATE=0.45, and (B) an increased infection time INF_TIME=6 during which the disease can be transmitted. Both changes result in disease persistence.

6.3 Exercises

6.1 Calculate the average and variance in the metapopulation code's insect populations, and plot these quantities as a function of the growth rate r. Can you make an analytic prediction?

6.2 Modify the metapopulation problem by giving patches a switching probability governing the transition between occupiable and unoccupiable each time step. Assume insects perish if the cell they migrate to is unoccupiable. How does the mean insect density vary with the switching probability?

6.3 Image the metapopulation dynamics using a PostScript file.

6.4 Write down a set of ordinary differential equations for an SEIR model (ignore the time delays). Solve for the equilibrium densities and compare them with results from a spatially mixed simulation. Do the two models' predictions match?

6.4 Projects

6.1 **Traffic Lights.** Convert the stochastic immigration–emigration model of chapter 2 into one where the emigration has an *off–on* cycle like a red–green traffic light. Have the immigration occur continuously through time. Examine how the period of the cycle time affects the average line length (if it does at all). Make sure the emigration rate increases during the on time to account for the zero emigration rate during the off time to give a temporally averaged emigration rate independent of cycle time in your comparison. What are the analytic expectations?

6.2 **Highway Driving.** This problem examines how a distribution of driving speeds affects throughput of traffic on a one-way one-lane highway. Use a highway of length L; cars of length D enter at one end with probability α per time step. Adjust these numbers to have around 100 cars on the highway. Initially assume that all cars move a distance v during a time step – this is a code-check because all cars should take L/v time steps to pass through the highway, and they should not bump into one another. Make a PostScript image of the cars on the highway as a function of time. In this test case, the image should be slanted parallel lines. Next, when a car enters the highway, randomly assign its driver a speed between $v - \Delta v$ and

$v + \Delta v$. Fast drivers get stuck behind slow drivers (cars cannot get closer than D). Fix L, v, and D, then examine the dynamics when α and Δv are changed. Compare images and averages.

6.3 **Gauntlet of Traffic Lights.** Imagine the Highway Driving project (6.2) with N traffic lights distributed randomly along the highway. Motorists must stop at all red lights. When going through a green light, assume a car can only start moving once a space has opened up between it and the car in front of it. Once moving, a car moves at its usual speed. When exiting the last light (the Nth one), cars leave the system entirely. Have all lights use the same cycle time, but with random phases. How does the average transit time vary with the number of lights?

6.4 **Punctuated Equilibrium.** Here is a simple model of macroevolution created by Sneppen *et al.* (1995). Use a one-dimensional lattice of 200 sites with wrapped ends, and for each site draw a random number between $B_0 > 0$ and 1. Call this number B_i for site i. Each site purportedly represents a species and B_i is something like the species' "fitness." First choose one site at random. For that site and its two neighbor sites draw new random numbers between 0 and 1 representing a new fitness value. Next, find the site with the smallest $B_i < B_0$ and randomly change its and its two neighbors' values. Repeat until there are no $B_i < B_0$. In the Sneppen *et al.* (1995) model, this change represents an evolutionary punctuation and the number of species involved in the event, s, is the punctuation size. What is the frequency distribution of punctuation sizes? How does this distribution change as a function of B_0? What is the average punctuation size and variance as a function of B_0?

6.5 **Pollinator Dynamics.** Imagine a one-dimensional line of F flowers, each flower having an initial amount of pollen A_0. Nectar-collecting pollinators enter at either end with rate η, continue through the line without turning, and exit at the opposite end. Also, assume they enter with no pollen on their bodies. A pollinator visits each flower with probability ν. If it visits a flower, then two things happen. First, the pollinator picks up a random fraction, $0 < \sigma < \sigma_{max}$, of the anthers' pollen. Second, a random fraction, $0 < \rho < \rho_{max}$, of the pollinator's pollen pool is deposited onto the stigma. If the pollinator doesn't visit the flower, then during the flight to the next flower in line it grooms away a random fraction, $0 < \beta < \beta_{max}$, of its pollen pool. What is the average pollen load per pollinator upon exit as a function of time? What are the anther and stigma pollen counts on each of the F flowers as a function of time?

6.6 **Prey–Predator Dynamics.** Consider a one-dimensional habitat modeled by an array of L sites. Each site can hold at most two prey and two predators. During a unit time step the main events are prey growth, predation and predator reproduction, predator death, and movement. At each cell occupied by a single prey individual, the prey reproduces a second individual with probability α during a time step. In cells occupied by both prey and predators, each predator has a probability β of attacking and consuming each prey individual. For each successful attack a new predator is produced with probability ϵ, however, the offspring dies if there is no predator vacancy in the cell. Predators perish with probability μ during each time step. Finally, individuals can move to a neighbor cell if the cell has a vacancy. The above description defines the spatial model. A mixed model replaces the local movement rule with one where each individual chooses a new location from all L sites each time step. Come up with *good* parameter values to compare the local and mixed models. What are the mixed model's dynamical equations?

6.7 **Additive genes.** Define a diploid individual having a series of L loci on homologous chromosomes. Each locus has A alleles. Maintain a population of fixed size, N, during the simulation. Initialize the population with random alleles. Individual fitness is additive; for each locus having allele 1 on both chromosomes, the individual scores a point. Total points is an individual's fitness. Score each individual's fitness. For each offspring of the next generation, choose two parents (with replacement [disallow selfing]) according to relative fitness. Each parent contributes one haploid gamete, produced as follows. Randomly choose one of *mom's* chromosomes. Copy the first allele to the gamete. Inaccurate copying occurs at each locus with mutation probability μ, which instead inserts a random chosen allele from the set of A alleles. With crossover probability χ, switch chromosomes. Copy/mutate the next allele and continue this process until all loci have been copied. Then repeat for *dad's* gamete. When all N individuals have been produced, measure each individual's fitness and produce the next generation. How does average fitness change with time? How do the results depend on A, L, N, χ, and μ?

6.8 **Food-chain model.** Here is a basic food-chain model

$$\begin{aligned} n_0(t+\Delta t) &= n_0(t) + \alpha \Delta t n_0(t)\left(1 - \frac{n_0(t)}{K}\right) \\ &\quad - \beta \Delta t n_0(t) n_1(t) \\ n_i(t+\Delta t) &= n_i(t) + \epsilon\beta\Delta t n_{i-1}(t) n_i(t) \\ &\quad - \beta \Delta t n_i(t) n_{i+1}(t) - \mu \Delta t n_i(t) \end{aligned}$$

Species n_0 (measured in units of biomass) is the primary producer, n_1 is the herbivore, etc. α is the resource growth rate, K is the resource carrying capacity, β is the rate one species is consumed by the species above it on the food chain, ϵ is the conversion efficiency of consumed biomass into produced biomass, and μ is the loss rate. Choose reasonable values for the parameters and something small for Δt. Start with the resource at its carrying capacity. Introduce a very small amount of n_1 and let the n_0–n_1 system equilibrate. If a population becomes negative, set it to zero. Then introduce a small amount of n_2 and let the new n_0–n_1–n_2 system equilibrate. How long a food-chain can you build? How important is Δt? Analytic predictions?

6.9 **Food-chain in a stream model.** This project uses the equations of the food-chain project (6.8). Imagine a small lake containing algae, the primary producer in our modeled aquatic system. Out of this lake runs a stream, which subsequently flows into another lake. Excluding the source lake, there are L such downstream lakes. Downstream from the Lth lake is a huge reservoir. A fraction $\rho\Delta t$ of the resource in one lake flows into the downstream lake each time step. The source lake's resource is maintained at K, so, for example, each time step an amount of resource $\rho \Delta t K$ enters the first downstream lake. Start all lakes at resource K. Start lake L with some small amount of consumer n_1. The reservoir maintains constant populations of all species $n_1 = 1$ to $n_N = 1$. Suppose that all consumers can move upstream and downstream equally well, such that a fraction $\mu\Delta t$ moves out of each lake, half going upstream, half going downstream. For example, each time step $\mu\Delta t/2$ of each species moves into the Lth lake from the reservoir. What is the food-chain length as a function of distance downstream?

7

Foraging Model

Programming components covered in this chapter:
• structures • root finding algorithm • clock-supplied seeds

Previous chapters examined stochastic processes operating between generations – birth and death processes – whereas this chapter focusses on interactions operating over a much shorter time scale. Specifically, this chapter begins with an examination of foraging by a single individual on a continuously renewing resource distributed over a collection of distinct patches (Possingham 1989, Possingham and Houston 1990, Nisbet *et al.* 1997). The forager's set of rules is very limited: Once in a patch, it stays for a fixed time, then randomly chooses a new patch from the full collection of patches. Time is incremented in discrete steps, during which patch resource values (assumed to represent biomass, energy, or similar resource variable) increase logistically, and the forager removes a fixed fraction of its patch's resource. Foraging does not come free – each interpatch transit inflicts a cost, measured in the resource currency, and an additional constant metabolic cost per unit time is imposed. Transit costs represent the energetic escapade of the forager leaving a patch, moving around the system, then settling onto another resource patch. The issues examined here involve resource dynamics within this collection of patches, and optimal foraging behavior on the consumer's part (e.g., Stephens and Krebs 1986, Krebs and Kacelnik 1991). A classic set of experiments, beginning with Milinski (1979), examined the foraging behavior of three-spined sticklebacks,[1] and a classic, biologically focussed and tested foraging model is that by Davies and Houston (1981)

[1] A species of fish, *Gasterosteus aculeatus*.

used to understand territorial defense by pied wagtails[2] against satellite consumers.

After a detailed examination of single-consumer models, I extend the foraging simulation into a spatial habitat containing multiple consumers. A biologically realistic alteration of the foraging rules, a trivial change in the simulation code, produces consumer grouping mediated only by resource interactions (Keller and Segel 1971, Gueron and Liron 1989, Wilson and Richards in press).

Several programs demonstrating a variety of computational features are examined in this chapter. After examining a deterministic model, I discuss an individual-based simulation introducing a new programming concept, *structures*, that gives C much of its power. I think of a structure as a user-defined variable containing many components. Structures are ideal for individual-based simulations because each individual is defined as a single element of an array of structure variables. Each of these elements can contain many variables, arrays, and even other structures and pointers to structures. Efficient use collects all aspects of a single individual into one structured unit.

7.1 Simplistic Analytic Foraging Model

The simplest description of the foraging model is the first of two analytic models that I cover. This section's model, closely related to Possingham and Houston's (1990) foraging model, provides a clear understanding of the system, unclouded by lots of mathematically picky details. Although it represents a crude deterministic approximation to the full problem, we'll see that, as crude as it is, it does a great job of characterizing the simulation results.

Imagine a system of N resource patches and a single consumer utilizing the patches. Clearly the consumer can only be in one place at a time, so $N - 1$ patches are always unoccupied by a consumer and there, happily, the resource grows. Let us assume that for a patch without a forager the resource r_i on patch i grows according to the logistic growth equation

$$\frac{dr_i}{dt} = \alpha r_i \left(1 - \frac{r_i}{K}\right), \tag{7.1}$$

where α is the resource growth rate and K is the patch carrying capacity. Our consumer has a very simple foraging behavior – it stays in a patch consuming resource at a constant fractional rate, β, for a time T before

[2] A species of bird, *Motacilla alba*.

moving on to a new patch. Resource dynamics for the forager-occupied patch (the oth patch) obeys the equation

$$\frac{dr_o}{dt} = \alpha r_o\left(1 - \frac{r_o}{K}\right) - \beta r_o. \tag{7.2}$$

How does the entire system's average resource density behave? Ignoring lots of details, we might hope that a crude average over all patches – one patch with a forager and $N-1$ identical patches without – provides a good approximation for the average patch dynamics. This approximation gives

$$\begin{aligned} \frac{dr}{dt} &= \frac{N-1}{N}\left[\alpha r\left(1 - \frac{r}{K}\right)\right] + \frac{1}{N}\left[\alpha r\left(1 - \frac{r}{K}\right) - \frac{\beta}{N} r\right] \\ &= \alpha r\left(1 - \frac{r}{K}\right) - \frac{\beta}{N} r, \end{aligned} \tag{7.3}$$

for the dynamics of the average patch resource r. The equilibrium resource density, r^*, as determined from $dr/dt = 0$, is

$$r^* = \left(1 - \frac{\beta}{\alpha N}\right) K. \tag{7.4}$$

If $N \to \infty$, then the forager's effect on the system becomes negligible and $r^* \to K$ as expected. On the other hand, if the per patch foraging rate, β/N, exceeds the patch growth rate, α, then the resource declines to $r^* = 0$ (a biologically realistic equilibrium of equation (7.3), unlike a negative resource density).

What is the situation from the consumer's perspective? One important measure is the consumer's net energy intake rate, $R(T)$, given the assumption that it stays on each patch for a time T. Continuing on in the spirit of a crude approximation, we can assume that the forager always lands on a patch initially at the equilibrium resource density r^*, which ignores all variance in patch resource densities. If so, how much resource does it consume? Net consumption from a patch, $g(T)$, is the time integral of the forager's instantaneous consumption rate minus its metabolic costs

$$g(T) = \int_0^T \beta r_o(t)dt - (\chi_t + \chi_m T). \tag{7.5}$$

Two costs are imposed: χ_t is the cost of traveling between patches, and χ_m is a constant metabolic rate. Assuming that a consumer is motivated to maintain a positive intake rate (by threat of death), a traveling cost results in a finite residence time, preventing the forager from moving quickly between patches and skimming a little bit of resource off each one. At the other limit, when the consumption rate exceeds the resource reproduction rate, a

Table 7.1. Parameter values used in the foraging model

Parameter	Value
α	0.05
K	1.0
β	0.1
χ_t	0.1
χ_m	0.01

metabolic cost forces the forager to move between patches rather than sit on one patch forever.

Determining the forager's net consumption demands solving equation (7.5), which requires solving $r_o(t)$ from equation (7.2). Rearranging and integrating

$$\int_{r^*}^{r_o(t)} \frac{dr_o}{(\alpha-\beta)r_o - \frac{\alpha}{K}r_o^2} = \int_0^t dt, \tag{7.6}$$

will yield $r_o(t)$, the amount of resource left on the patch after a forager has been on the patch for a time t given that the initial resource density was r^*. I gave the results of this integral earlier for the density-dependent birth–death model (3.9). In terms of the present variables, the solution is

$$r_o(t) = \frac{r^*K(\beta-\alpha)}{(\alpha r^* + K(\beta-\alpha))e^{(\beta-\alpha)t} - \alpha r^*}. \tag{7.7}$$

Placing $r_o(t)$ into equation (7.5) and integrating again gives

$$\begin{aligned} g(T) &= \frac{\beta K}{\alpha}\left(\ln\left(\frac{(\alpha r^* + K(\beta-\alpha))e^{(\beta-\alpha)T} - \alpha r^*}{K(\beta-\alpha)}\right) - (\beta-\alpha)T\right) \\ &\qquad - (\chi_t + \chi_m T). \end{aligned} \tag{7.8}$$

Finally, the net foraging rate, $R(T)$, is the net resource obtained, $g(T)$, divided by the total time spent on the patch, T

$$R(T) = \frac{g(T)}{T}. \tag{7.9}$$

Figure 7.1 displays the results from this approximate analytic model using the parameter values shown in table 7.1. As seen from the results, the optimal residency time is somewhere between 5 and 20 time units, dependent on the number of patches in the system.

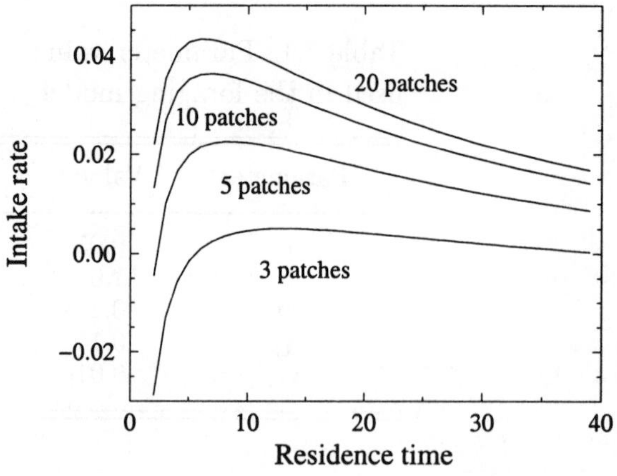

Figure 7.1. Analytic results for the intake rate, $R(T)$, for a single consumer foraging in multipatch systems (see equation (7.9)). Residence time, T, refers to how long the consumer stays on a patch to forage before leaving for another patch.

7.2 Optimal Residence Time

If you were a consumer, how long would you stay on a patch? Here I address the question by maximizing the foraging rate on a patch (assuming behavioral or evolutionary mechanisms are at play) by optimizing the time spent on each patch (Yamamura and Tsuji 1987).[3] Analytically, the foraging rate, $R(T) = g(T)/T$, is differentiated and set to zero

$$\left.\frac{dR(T)}{dT}\right|_{T=T^*} = -\frac{g(T^*)}{T^{*2}} + \frac{g'(T^*)}{T^*} = 0. \tag{7.10}$$

where $g'(T) = dg(T)/dT$. In general you should make sure that the second derivative is negative, $R''(T) < 0$, such that the solution to equation (7.10) is a maximum, not a minimum. Equation (7.10) now leads to

$$g'(T^*) = \frac{g(T^*)}{T^*} \tag{7.11}$$

as the solution for the optimal time T^* spent foraging on a patch. This expression is a statement of the marginal value theorem – the time to leave a patch is when the expected rate of return, $g'(T)$, drops to the average rate of return obtained by moving to a new patch (Charnov 1976).[4]

[3] My approach of maximizing the time spent on a patch disregards selection-altering foraging behaviors in more fundamental ways (Abrams 1982).

[4] A good introductory text discussing these concepts is Bulmer's *Theoretical Evolutionary Ecology* (1994). Another good book is Charnov (1982) on the importance of the marginal value theorem in behavioral ecology and evolutionary systems.

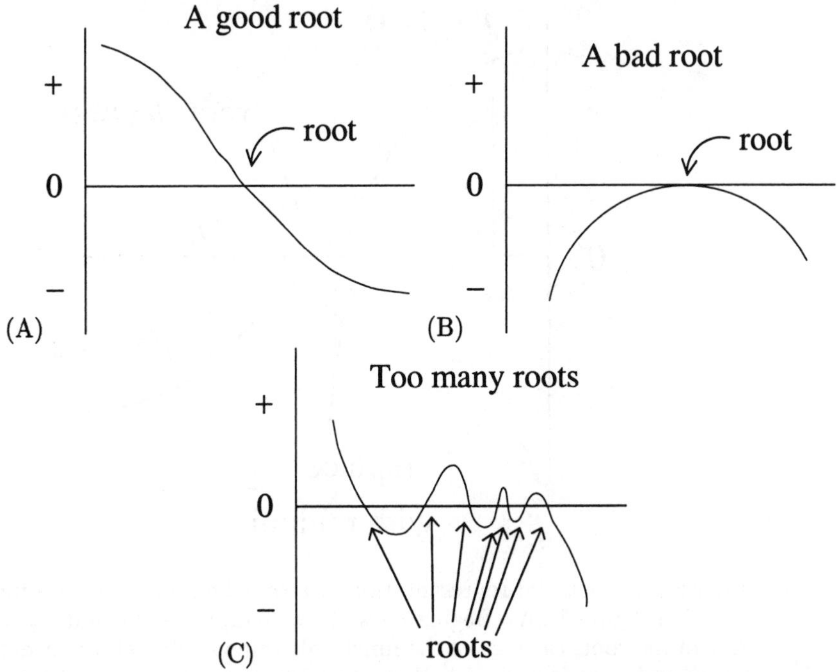

Figure 7.2. Caricatures of (A) good and (B,C) bad root finding scenarios. In (A) the good function whose root is to be located has positive values on one side of the root and negative values on the other. (B) In contrast, one bad function has values of the same sign on both sides, and (C) another bad function has so many roots that isolating any one of them is difficult.

Using the marginal value theorem, equation (7.11), with the expression for $g(T)$, equation (7.8), we can determine the optimal T^*. It would be nice if we could get an explicit expression for T^* as a function of all the model parameters, but the resulting expression does not allow T^* to be isolated on one side of an equation by algebraic rearrangements.[5] We are left with an implicit solution for T^*, meaning we must obtain a numerical estimate involving a procedure called root finding. Of course one could use "canned" algorithms provided by most numerical analysis programs, but that would take the fun out of life, wouldn't it?

7.2.1 Root Finding Algorithm

Determining the optimal residence time requires finding the value of T such that $g'(T) - g(T)/T = 0$ (equation (7.11) rewritten). This particular value is called the root of the expression. In uncomplicated cases pictured in figure

[5] A simple example of such a transcendental equation is $x = e^{ax}$; try solving for x.

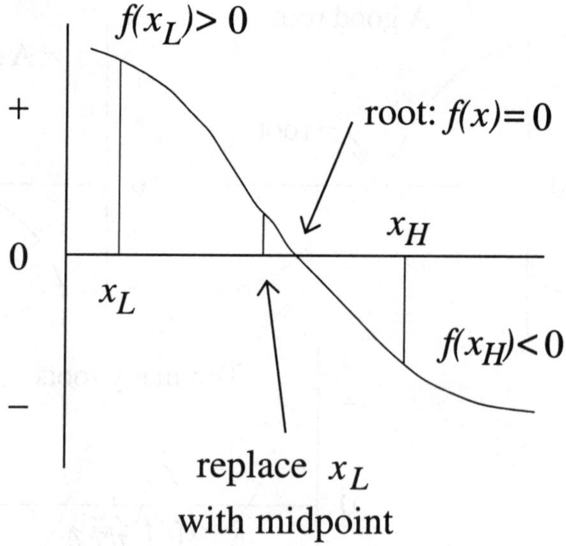

Figure 7.3. Pictorial representation of a root finding algorithm for a good function (figure 7.2A). Beginning with two endpoints, x_L and x_H, on either side of the root, the sign of the function's value at the midpoint determines which endpoint is replaced. Repeating this procedure many times zeros in on the root.

7.2A, there is a single root, and for values of T on either side of the root, the expression $g'(T) - g(T)/T$ is nonzero and has opposite signs. Designing an algorithm that zooms in on the root, given two starting points on either side of the root, is then relatively straightforward as shown below. It becomes a bit of an art when dealing with the more general (and pathological) root finding problems shown in figure 7.2B,C.

Figure 7.3 demonstrates the essential features of the root finding procedure implemented below. First, extreme limits for the upper and lower bounds are estimated, and the only critically important aspect is that the initial range contains the root. The procedure then replaces with the midpoint the extreme value whose root function evaluates to the same sign as the midpoint. A single iteration of this procedure reduces the root-enclosing range by one-half. Repeat until the range is an acceptably small value.

7.2.2 Code

```
#include <stdlib.h>
#include <stdio.h>
#include <math.h>
```

```
     #define   ALPHA     0.05
     #define   KK        1.0
     #define   BETA      0.1
     #define   MOVECOST  0.1
     #define   TIMECOST  0.01

     double    alpha, beta, k, cost1, cost2, rstar;
     double    RootFn(double);
     double    tgprime(double);
     double    ggg(double);

     int main(void)
     {
          int       step;
          double    lowt, midt, hight, lowf, midf, highf, n, nfac;

          alpha = ALPHA; beta = BETA; k = KK;
          cost1 = MOVECOST; cost2 = TIMECOST;

          for(n=3.0,nfac=1.0;n<1000.0;n+=nfac*0.1,nfac*=2.0)
          {
               rstar = (alpha-beta/n)*k/alpha;
1.a)           lowt = 0.0001; hight = 1000.0;
               lowf = RootFn(lowt); highf = RootFn(hight);

               for(step=0;step<50;step++)
               {
1.b)                midt = (lowt + hight)/2.0;
                    midf = RootFn(midt);

1.c)                if(lowf*midf<0.0)        hight = midt;
                    else if(highf*midf<0.0) lowt = midt;
                    else break;
               }
               printf("%g %g \n",midt,ggg(midt));
          }
          return (0);
     }

2.a) double RootFn(double x)
     {
          return( gprime(x) - ggg(x)/x );
     }

2.b) double gprime(double x)
     {
          return( (ggg(1.001*x)-ggg(x))/(0.001*x) );
     }
```

```
2.c) double ggg(double x)
     {
          double mess;

          mess = ( (alpha*rstar+k*(beta-alpha))*exp((beta-alpha)*x)
                    -alpha*rstar) / k / (beta-alpha);
          return( (beta*k/alpha)*(log( mess ) - (beta-alpha)*x)
                    - (cost1+cost2*x) );
     }
```

7.2.3 Code Details

Line 1: These lines contain the heart of the root finding algorithm. At lines 1.a extreme limits for the upper and lower bounds are estimated for the root, such that evaluations of `RootFn()` have opposite signs. We could perform tests on the code to make sure this choice was made correctly; an alternative is just recognizing spurious results when the program runs. Calculation of the range's midpoint occurs at line 1.b, along with its value for `RootFn()` on the following line. Within the `if-else` construction at line 1.c, the extreme value whose `RootFn()` value has the same sign as the midpoint's is thrown away, and the midpoint replaces it as the extreme value. Repeating this process 50 times reduces the range by a factor of $0.5^{50} = 10^{-17}$, which is an accuracy well within our demands.

Line 2: At line 2.a, `RootFn()` represents the expression $g'(T) - g(T)/T$, the function whose value is zero at the root value of T and nonzero elsewhere (I use `x` for the local variable representing T). `RootFn()` itself calls two functions, `gprime()` and `ggg()`. The first represents $dg(T)/dT$, which I calculate numerically in the function at line 2.b, using two calls to `ggg()` with closely spaced values for T. A numerical approach like this would be used in complicated problems having no analytic representation; however, in this case, we could differentiate equation (7.8) directly. The second function, `ggg()`, is just the function $g(T)$, equation (7.8).

7.2.4 Optimal Residence Time Results

Figure 7.4 adds a curve depicting the intake rates obtained at the optimal residence times for a large range of N-patch systems. Sure enough, the calculation matches the peaks of the intake rate curves.

Intuitively, T^* decreases with increasing habitat size because the forager

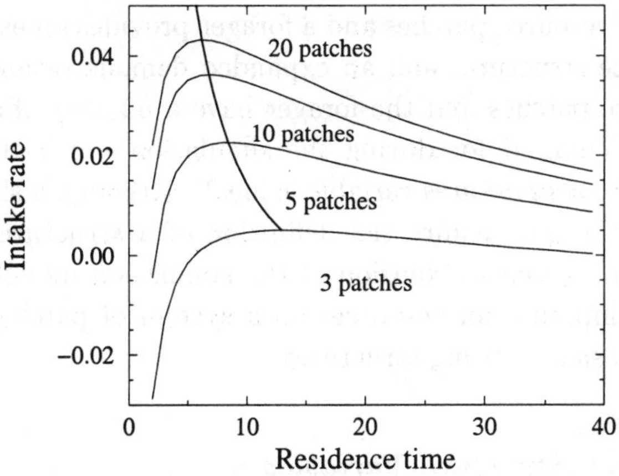

Figure 7.4. Results for the optimal residence times plotted on top of the analytic intake rates shown in figure 7.1.

should leave a patch when it can get a better return elsewhere. With an increasing number of patches, it is increasingly assured of moving to a new patch completely filled with resource. For an extremely large number of patches, the forager should only stay for an exceedingly small amount of time after it recoups its transit costs (Charnov 1976).

Should there be an optimum? No matter what the case, a consumer should stay long enough on a patch to recoup its transit costs (if the patches are full) or reduce its expenditures (if the patches are empty). Thus we anticipate the increase in intake rate very close to $T = 0$. What about for high values of T? There is likely to be an optimum intake rate[6] whenever $r^* > 0$ (equation (7.4)). For the parameter values used here, $r^* = 0$ for $N = 2$ patches, so that situation is unclear. However, for $N \geq 3$, staying forever on a single patch, consuming its resource to zero and gaining no additional benefit is a bad strategy because the other patches will eventually become resource filled. Hence the intake rate must also decrease with increasing T when T is very large, meaning there must be an optimum somewhere between 0 and ∞.

7.3 Forager Simulation Model

I now present a simulation model of a single forager in a collection of patches. The simulation loops through time, moving the consumer from patch to patch removing resource, and replenishing the patches' resources. Modeling

[6] Although the optimum might be negative if the forager's costs are too high.

a system of resource patches and a forager provides an excellent opportunity to introduce structures and an expanded demonstration of pointers. Both the resource patches and the forager have a number of state variables that must be accounted for during the simulation – collecting these variables together into a *structures variable* is ideal. Although a simulation of a single forager might not require the definition of a structure for organizational purposes, an obvious extension of the simulation model is a population of foragers competing for resources on a system of patches. Such extensions are made easier by using structures.

7.3.1 Structure Variables

C allows for the collection of related variables (of arbitrary types) into a single entity. These entities are called structures, and the variables collected together are called the structure's members and are stored together in memory. In the following program, two structures will be defined for patches and consumer, respectively. For example, an array of patch structures is declared by the following lines of code

```
struct patch_def
{
      double    k;
      double    alpha;
      double    size;
} patches[NPATCH];
```

The word `struct` warns the compiler that a structure entity is coming up. The next word, `patch_def`, is called the structure *tag*, a name identifying a specific structure definition. The tag does not declare an actual structure variable, rather it only identifies the template, or format, of the structure. Declarations within the braces of lines 1.a and 1.b define the structure members. Structure variables are declared after the closing braces: `patches[NPATCH]` is declared as an array of patch structures, for which computer memory is set aside.

Figure 7.5 shows schematically how the array of `patch_def` structures, with its structure members, is stored in memory. The period operator (also called the *member access operator*) marks the separation between the structure name on the left-hand side and the structure member on the right-hand side. The address of a particular structure variable is accessed in the usual way using the address operator `&`. For example, in figure 7.5 the address of structure variable `patches[2]` would be accessed by `&patches[2]` yielding the address `00A49C10`. The address of a particular member could also be

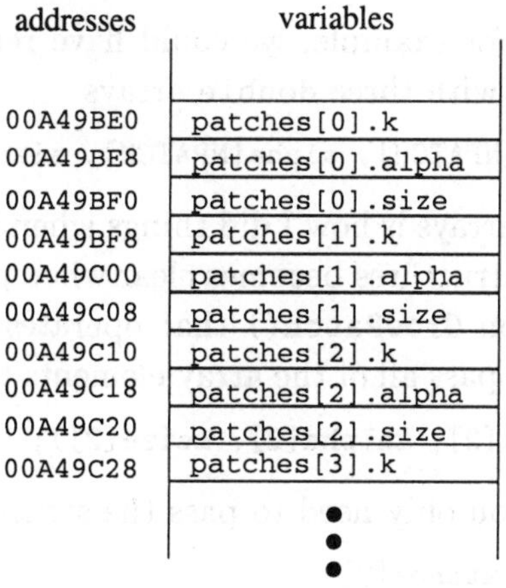

Figure 7.5. Structure variables collect related variables (of arbitrary type) into contiguous chunks of memory. The address of a structure variable or one of its members is accessed by, for example, **&patches[2]** or **&(patches[2].size)**, respectively. A pointer to a structure variable can also be a valid structure member.

obtained, for example **&(patches[2].size)** yields the address 00A49C20. A pointer to a structure can be declared, for example

```
struct patch_def *patpnter;
```

and one can place the address of a structure into a structure pointer, such as the line

```
patpnter = &patches[2];
```

Given a structure or a pointer to a structure, structure members can be accessed in several ways, for example

```
patches[2].k = 1.0;
(*patpnter).k = 1.0;
patpnter->k = 1.0;
```

The first line is the most direct method of accessing the member. Given the above assignment of **patpnter**, the second line turns the structure pointer into a structure variable using the indirection operator, *****, then uses the period operator to access the member **k**. The quickest and cleanest way of accessing structure members uses the arrow operator **->**, shown on the third line.

Why would you want to use structures? Again, structures allow the collection of related pieces of information into a single entity. You certainly don't need to use them; everything you can do with structures you can do some other way. For example, we could have replaced the structure array `patches[NPATCH]` with three `double` arrays

```
double    k[NPATCH], alpha[NPATCH], size[NPATCH];
```

Defining multiple arrays is how I did things when I programmed in Fortran.[7] The advantage of structures becomes clear when passing values to functions. Consider a function `GrowPatch()` that operates on a single patch. Using arrays you need to pass all of the array elements to the function directly, i.e.

```
GrowPatch(&k[2], &alpha[2], &size[2]);
```

Using structures you only need to pass the structure pointer

```
GrowPatch(&patches[2]);
```

If, for some reason, you want to add a new patch variable and you use arrays, then you need to define a new array and modify all the function calls everywhere in your code to accept the added array's elements. If you use structures, you simply add a member to the structure definition and modify the appropriate functions that use it. No function calls need to be altered since the entire structure's pointer is passed. These arguments become much more convincing if you have structures with dozens of members all of arbitrary data types.

7.3.2 Code

```
     #include  <stdlib.h>
     #include  <stdio.h>
     #include  <math.h>
2.b) #include  <time.h>

     #define   MAXTIME   (500)
     #define   NPATCH         3
     #define   ALPHA          0.05
     #define   KK        1.0
     #define   CONFRAC        0.1
     #define   MOVECOST  0.1
     #define   TIMECOST  0.01
     #define   STAYTIME  10

     struct patch_def
     {
          double    k;
```

[7] In this specific example you could define a two-dimensional array to the same effect as a structure, but, more generally, structures allow you to mix data types.

```
          double    alpha;
          double    size;
     } patches[NPATCH];

1.a) struct con_def
     {
          double    beta;
          int  staytime;
          struct patch_def *patch;
          double    time;
          double    lunch;
          double    inrate;
     } consumer;

     void InitSeed(void);
     void InitCon(struct con_def *);
     void InitPatch(struct patch_def *);
     void GrowPatch(struct patch_def *);
     void Feed(struct con_def *);
     void Move(struct con_def *);

     int main(void)
     {
          int  ipat, ttt;

          InitSeed();
1.b)      InitCon(&consumer);
          for(ipat=0;ipat<NPATCH;ipat++)
               InitPatch(&patches[ipat]);

          for(ttt=0;ttt<MAXTIME;ttt++)
          {
               for(ipat=0;ipat<NPATCH;ipat++)
                    GrowPatch(&patches[ipat]);

               Feed(&consumer);
               Move(&consumer);

               printf(" %d ", ttt);
               for(ipat=0;ipat<NPATCH;ipat++)
                    printf(" %f ",patches[ipat].size);
               printf(" %f \n", consumer.inrate);
          }
          return (0);
     }

2.a) void InitSeed(void)
     {
          int  seed;
2.c)      time_t    nowtime;
```

```
          struct    tm *preztime;

2.d)      time(&nowtime);
2.e)      preztime = localtime(&nowtime);
2.f)      seed = (int)((preztime->tm_sec+1)*(preztime->tm_min+1)*
               (preztime->tm_hour+1)*(preztime->tm_year)
                    *(preztime->tm_year));
2.g)      if(seed%2==0) seed++;
          srand48(seed);
     }

1.c) void InitCon(struct con_def *con)
     {
1.d)      (*con).beta = CONFRAC;
1.e)      con->staytime = STAYTIME;
          con->patch = &patches[0];
          con->time = 0;
          con->lunch = -MOVECOST;
          con->inrate = 0.0;
     }

     void InitPatch(struct patch_def *pat)
     {
          pat->k = KK;
          pat->alpha = ALPHA;
          pat->size = KK;
     }

     void GrowPatch(struct patch_def *pat)
     {
          pat->size += pat->alpha * pat->size
                    * (1.0-pat->size/pat->k);
          if(pat->size>pat->k) pat->size = pat->k;
     }

     void Feed(struct con_def *con)
     {
          double eaten;
          eaten = con->beta * con->patch->size;
          con->patch->size -= eaten;
          con->lunch += eaten;
          con->time += 1.0;
     }

     void Move(struct con_def *con)
     {
          if(con->time>=con->staytime)
          {
               con->time = 0.0;
               con->lunch -= TIMECOST*con->time;
```

```
            con->inrate = con->lunch/con->time;
            con->lunch = -MOVECOST;
            con->patch = &patches[(int)(drand48()*NPATCH)];
        }
    }
```

7.3.3 Code Details

Line 1: A complicated structure variable is defined on line 1.a, having members of types `double` and `int`, and pointers to other structures. If needed, you could also include self-referential members that are pointers to the defined structure type. In this case I allocate space only for a single consumer structure, `consumer`. Line 1.b passes the address of the consumer structure to `InitCon()`, hence `con`, defined as a pointer on line 1.c, points to the globally defined structure variable `consumer`. As discussed above, given a correctly assigned structure pointer, structure members can be accessed in two equivalent ways, shown on lines 1.d and 1.e. Notice that passing a single pointer to the structure allows the function access to the entire set of variables defined within the structure. Without structures, each variable would have to be passed explicitly to a function.

Line 2: Clock-supplied seeds. `InitSeed()` at line 2.a gives a different random number generator seed every time the program is executed. It makes calls to the CPU's clock and extracts rapidly (and slowly) varying numbers. You need to call the `time.h` header file (line 2.b) that defines the structures `time_t` and `tm` at 2.c. `InitSeed` first calls the function `time` at line 2.d, which returns the structure `nowtime` with appropriate values set. A second call to `localtime()` at line 2.e gives the numbers used for the random seed. The structure members of `tm` are all of type `int`: `tm_sec` gives the number of seconds past the minute; `tm_min` gives the number of minutes past the hour; `tm_hour` gives the number of hours since midnight; and `tm_year` gives the number of years since either 1900 or 1970 (I've seen both used.). Multiplying these numbers together at line 2.f gives us a large number that varies across two orders of magnitude over one-minute time scales. I cast this result into an `int` in case it gets too big. Hence if two runs are started even a few seconds apart, their random number streams likely will be very different. I once heard that random number generator seeds should be odd numbers (line 2.g).

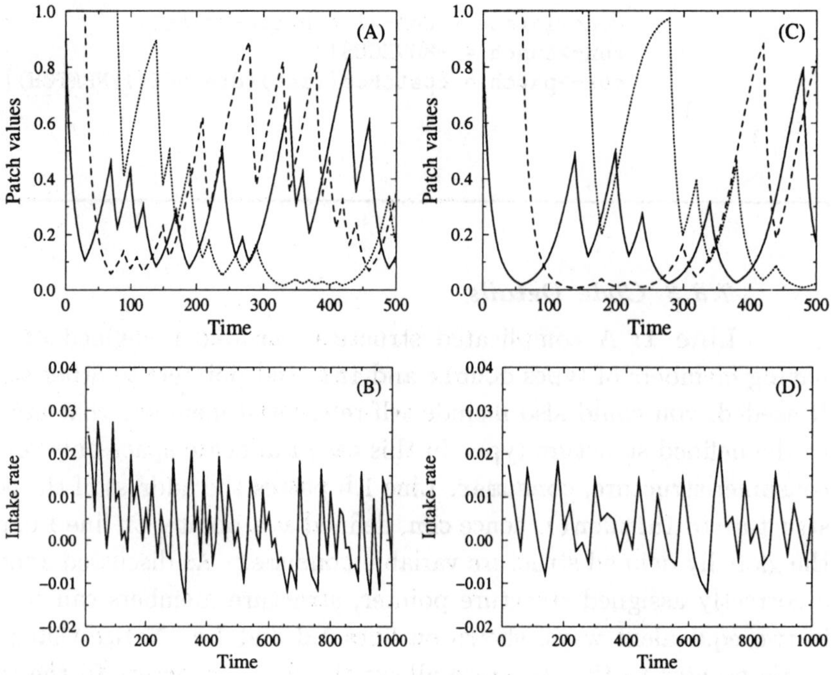

Figure 7.6. Simulation results of a consumer foraging in a system of three resource patches. The consumer has patch residence times (A,B) $T = 10$ and (C,D) $T = 20$. (A,C) Plots of the patch resource values during the simulation. Sharp drops in resource value correspond to the arrival of the consumer. (B,D) Consumer net intake averaged over the patch residence time.

General: Notice how clean the `main()` routine is, given how much is done here. All of the detailed work is placed into functions that do small parts of the problem, and information is passed efficiently using structures, leaving `main()` easily interpreted in terms of the logic of the biological problem at hand. Almost anyone can see what is happening in the program, or at least within `main()`. Clarity is important. One could certainly remove all of the functions, define arrays to hold data, and put everything into `main()`, but readability will go way down.

7.3.4 Simulation Results

Figure 7.6 presents results for the patch resource densities and the consumer's total intake from a simulation of a three-patch habitat. Parameter values are specified in table 7.1. As specified in the simulation code, the patch resource densities start at their carrying capacity. Figure 7.7

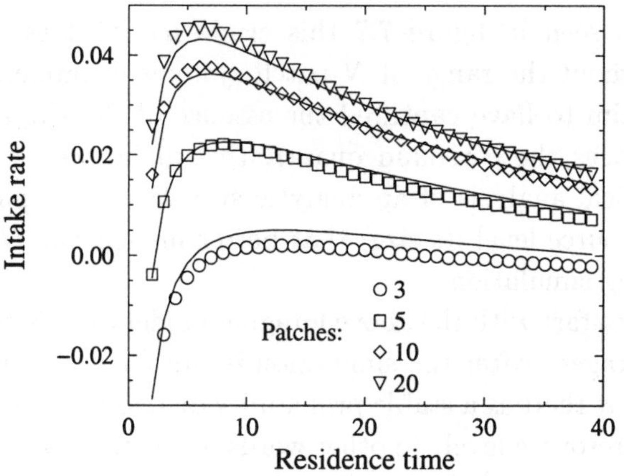

Figure 7.7. Comparisons between the simulated consumer's average net intake rate and the predictions of the simplistic foraging model (figure 7.1). Overall good agreement is observed between the results in spite of the simplicity of the one model and the stochasticity of the other.

examines the intake rate as a function of residency time T for several N-patch systems and compares them with the analytic predictions from the deterministic model (figure 7.1) of section 7.1. (These calculations required code modifications – a loop over residency times and a measurement, after transients have decayed away, of the consumer's intake rate. Intake rates are averaged for 10^6 simulation steps.)

Good agreement between the analytic and simulation models is seen for a large range of N. The agreement seems too good considering the crudeness of the analytic model's approximation. Namely, it assumes that all new patches the consumer moves to are at "equilibrium," but the range of initial patch resource densities the consumer sees in the simulation (marked by the heights of the peaks) is very broad. Apparently this variability has little influence on the mean intake rate the consumer experiences. However, being a bit more sticklish for agreement, the approximate model breaks down in a variety of places, especially for systems with only a few patches. This lack of agreement motivates a more complex analytic model, accounting for the distribution of patch resource sizes.

7.4 Deterministic Size-Structured Patch Model

A defining assumption made in the previous analytic model was that the forager always landed on a patch with initial resource density r^* (equation

(7.6)). As seen in figure 7.7 this crude model does a surprisingly good job throughout the range of N-patch systems examined. In many respects we can claim to have captured the essence of the simulation. However, in various places there remain quantitative differences. This section covers the numerical analysis of an analytic size-structured patch model (I call a patch's resource level its size) that better models the N-patch dynamics of the foraging simulation.

Again we start with the above foraging model with N resource patches and a single forager. After the simulation is run for a very long time, we might imagine that there is a stable probability distribution of finding patches at a particular resource level. In other words if one million simulations of a three-patch system are run in parallel, then at any instant (after equilibration) we could measure the resource level in all three million patches and plot a histogram of the number of patches found within small intervals of resource density. Dividing this histogram by the total number of patches yields a frequency distribution of resource densities. Repeating this measurement at another time should yield the same frequency histogram (within uncertainties). This distribution is the probability of finding a randomly chosen patch at a specific resource level, and once we have this patch resource distribution we can calculate the forager's expected intake rate.

I will use a discretized conveyor belt picture of a size-structured population of patches in deriving the analytic formulation (figure 7.8). The conceptual simplification breaks the continuous-size-structure into small units of size Δs. The distribution of patches at each of these discrete sizes is imagined as spread uniformly over a conveyor belt, and the total fraction of patches having size s is the volume of the block on the conveyor belt, $n(s,t)\Delta s$. The conveyor belt centered on size s moves with speed $g(s)$, where $g(s)$ is the growth rate of patches of size s, and represents how quickly patches move through a particular size class.

During a small time interval Δt, the conveyor belt at s moves through a distance $g(s)\Delta t$; likewise the conveyor belt at $s - \Delta s$ moves through a distance $g(s-\Delta s)\Delta t$. Since these are the only processes affecting the conveyor belt at s, its patch density updates according to the discretized equation

$$n(s,t+\Delta t)\Delta s = n(s,t)\Delta s + [g(s-\Delta s)\Delta t]n(s-\Delta s,t) - [g(s)\Delta t]n(s,t). \tag{7.12}$$

Essentially this equation treats the patch distribution as soil being moved along this series of conveyor belts with the most important aspect being that soil is neither created nor destroyed. Although we will compare equation (7.12) with simulation results, deriving the continuous-time, continuous-size

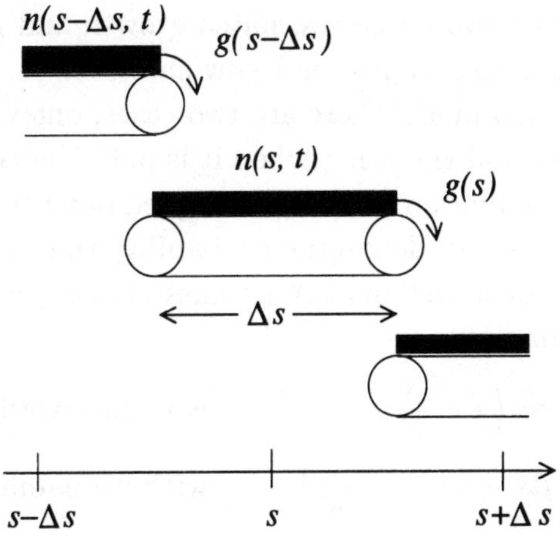

Figure 7.8. Conveyor belt picture of size-structured population model (after Gurney and Nisbet 1983). In a short time Δt, the middle conveyor belt, for example, turns through a distance $g(s)\Delta t$, depositing an amount of population $n(s,t)g(s)\Delta t$ onto the next conveyor belt up the line.

structured model from here is straightforward. Several terms in equation (7.12) can be Taylor series expanded, and assuming Δs and Δt are small enough to ignore terms proportional to $(\Delta s)^2\Delta t$, $\Delta s(\Delta t)^2$ and higher, we obtain

$$\left[n(s,t) + \frac{\partial n(s,t)}{\partial t}\Delta t\right]\Delta s = n(s,t)\Delta s + \left[\left[g(s) - \frac{\partial g(s)}{\partial s}\Delta s\right]\Delta t\right]\left[n(s,t) - \frac{\partial n(s,t)}{\partial s}\Delta s\right] - [g(s)\Delta t]n(s,t) \tag{7.13}$$

$$\frac{\partial n(s,t)}{\partial t}\Delta t\Delta s = -\frac{\partial g(s)}{\partial s}n(s,t)\Delta s\Delta t - g(s)\frac{\partial n(s,t)}{\partial s}\Delta s\Delta t, \tag{7.14}$$

to yield

$$\frac{\partial n(s,t)}{\partial t} = -\frac{\partial}{\partial s}[g(s)n(s,t)]. \tag{7.15}$$

Along with associated initial and boundary conditions, this equation represents the simplest possible size-structured population dynamics model. Similar discussions, but with more details, are given by Sinko and Streifer (1967), Gurney and Nisbet (1983), and Murray (1989). Equation (7.15) says, in words, that a temporal change in the frequency of patches of size

s is a result of the two processes, patch growth and grazing by the forager, embodied in the function $g(s)$ as I now show.

In the simulation model there are two cases, one when the patch is occupied by a forager and the other when it is not. Unoccupied patches grow in size according to equation (7.1) and occupied patches are depleted according to equation (7.2). If we define the probability that a forager is grazing on a patch, γ, which for a system of N patches is just $\gamma = 1/N$, then the growth and depletion functions are

$$g_e(s) = \alpha s \left(1 - \frac{s}{K}\right) \qquad \text{with probability} \quad 1-\gamma \tag{7.16a}$$

$$g_o(s) = \beta s - \alpha s \left(1 - \frac{s}{K}\right) \qquad \text{with probability} \quad \gamma \tag{7.16b}$$

for empty and occupied patches, respectively. Here I explicitly incorporated using a minus sign that, on an occupied patch, grazing exceeds growth. We then have an overall growth function, $g(s) = (1-\gamma)g_e(s) - \gamma g_o(s)$. Replacing this overall growth function into equation (7.15) gives

$$\frac{\partial n(s,t)}{\partial t} = -(1-\gamma)\frac{\partial}{\partial s}[g_e(s)n(s,t)] + \gamma\frac{\partial}{\partial s}[g_o(s)n(s,t)]. \tag{7.17}$$

Although continuous-time and continuous-size equations are cleaner and more mathematically appealing, one must usually implement a numerical solution. Numerical solution of equation (7.17) demands discretization, putting us right back to an equation something like equation (7.12). I implement a numerical integration method called upwind (and downwind) differencing (e.g., Strikwerda 1989) to examine the forager simulation results. More detailed and refined integration schemes and routines for structured population models are readily available (de Roos, Diekmann, and Metz 1992, de Roos 1996, Richards 1997). Discretizing equation (7.17) by setting

$$\frac{\partial n(s,t)}{\partial t} \approx \frac{n(s,t+\Delta t) - n(s,t)}{\Delta t} \tag{7.18a}$$

$$\frac{\partial}{\partial s}[g_e(s)n(s,t)] \approx \frac{g_e(s)n(s,t) - g_e(s-\Delta s)n(s-\Delta s,t)}{\Delta s} \tag{7.18b}$$

$$\frac{\partial}{\partial s}[g_o(s)n(s,t)] \approx \frac{g_o(s+\Delta s)n(s+\Delta s,t) - g_o(s)n(s,t)}{\Delta s}, \tag{7.18c}$$

yields our numerical scheme

$$n(s,t+\Delta t) = n(s,t)\left(1 - (1-\gamma)g_e(s)\frac{\Delta t}{\Delta s} + \gamma g_o(s)\frac{\Delta t}{\Delta s}\right)$$

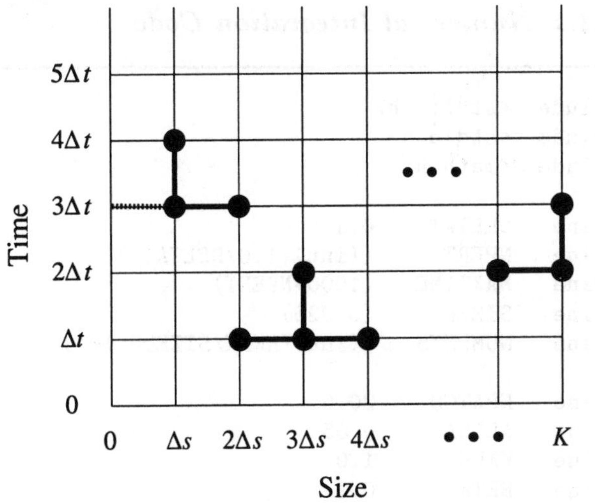

Figure 7.9. Schematic diagram outlining the numerical solution to the size-structured patch model used in the program of section 7.4.1. Upwind and downwind differencing is used to update the size distribution from earlier times (see equation (7.19)).

$$+ (1-\gamma)\frac{\Delta t}{\Delta s} g_e(s-\Delta s)n(s-\Delta s,t)$$
$$- \gamma\frac{\Delta t}{\Delta s} g_o(s+\Delta s)n(s+\Delta s,t), \tag{7.19}$$

easily translated into a computer program. A schematic of the algorithm is shown in figure 7.9. The algorithm requires at time $t=0$ a full specification of the values $n(\Delta s, 0)$, $n(2\Delta s, 0)$, and so on. It also requires boundary conditions for the left- and right-most nodes; I implement no-flux boundaries, meaning that the flow along the dashed line connecting to the $s=0$ cells is zero, and because we use logistic growth, there is no flow beyond $s=K$.

The scheme uses equation (7.19) to calculate $n(i\Delta s, (j+1)\Delta t)$ from values at three earlier nodes, along with the initial and boundary conditions. The essential idea behind the code is to begin with an initial size-structured distribution and update the distribution until it reaches a steady state. Once steady state is acheived we calculate the forager's expected intake rate given this distribution of patches. Generally, the numerical scheme will give a good solution as long as the Courant condition (e.g., Press *et al.* 1986), $|g_e\Delta t/\Delta s|, |g_o\Delta t/\Delta s| < 1$, is satisfied (see exercise 7.9).

7.4.1 Numerical Integration Code

```
    #include  <stdlib.h>
    #include  <stdio.h>
    #include  <math.h>

    #define   DELTAT     0.1
    #define   NPERT      ((int)(1.0/DELTAT))
    #define   MAXTIME    (1000*NPERT)
    #define   SIZEDEL    (0.025)
    #define   NUMPNTS    ((int)(KKKK/SIZEDEL)+1)

    #define   NPATCH     20.0
    #define   ALPHA      0.05
    #define   KKKK       1.0
    #define   BETA       0.1
    #define   MOVECOST   0.1
    #define   TIMECOST   0.01

    double    sizes1[NUMPNTS], sizes2[NUMPNTS],
              growth[NUMPNTS], graze[NUMPNTS],
              dubsize[NUMPNTS];
    double    grazerate;
    void      InitStuff(void);
    void      InitSizeDist(double *sizedist);
    void      UpdateSizeDist(double *, double *);
    double    Eaten(double, double *);
    double    grazed(double, double);

    int main(void)
    {
         int        ttt, size;
         double     staytime, aveeaten, *s1, *s2, *stmp;

         InitStuff();
              /* Determine the steady-state size distribution */
1.a)     s1 = sizes1; s2 = sizes2;
         InitSizeDist(s1);

         for(ttt=0;ttt<MAXTIME;ttt++)
         {
1.b)          UpdateSizeDist(s1,s2);
1.c)          stmp = s1; s1 = s2; s2 = stmp;
         }
                                        /* Calculate intake rates */
         for(staytime=2.0;staytime<40.0;staytime+=1.0)
         {
              aveeaten = Eaten(staytime,s1);
              printf(" %4.1f %f\n", staytime, aveeaten);
```

```
    }
    return (0);
}

void InitStuff(void)
{
    int       size;
    double    val;

    grazerate = 1.0/NPATCH;        /* one patch has a grazer */

    for(size=1;size<NUMPNTS;size++)
    {        /* tabulate growth/grazing functions */
        val = ((double)size)*(double)SIZEDEL;
2.a)    dubsize[size] = val;
2.b)    growth[size] = ALPHA*DELTAT*val
              *(1.0 - val/KKKK)/SIZEDEL;
2.c)    graze[size] = -(growth[size]
                - BETA*DELTAT*val/SIZEDEL);
    }
}

void InitSizeDist(double *sizedist)
{
    int  size;
    for(size=1;size<NUMPNTS-1;size++) sizedist[size] = 0.0;
    sizedist[NUMPNTS-1] = 1.0; /* start with largest R */
}

void UpdateSizeDist(double *old, double *new)
{
    int       size;
    double    norm;

    new[1] = old[1] * (1.0 - (1.0-grazerate)*growth[1])
            + grazerate*graze[2]*old[2];

    for(size=2;size<NUMPNTS-1;size++)
    {
        new[size] = old[size] *
            (1.0 - (1.0-grazerate)*growth[size]
                - grazerate*graze[size])
            + (1.0-grazerate)*growth[size-1]*old[size-1]
            + grazerate*graze[size+1]*old[size+1];
    }
    new[NUMPNTS-1] = old[NUMPNTS-1]
        * (1.0 - grazerate*graze[NUMPNTS-1])
        + (1.0-grazerate)*growth[NUMPNTS-2]*old[NUMPNTS-2];
}
```

```
double Eaten(double stime, double *sizedist)
{
     int        size;
     double     xxx;

     xxx = -MOVECOST-TIMECOST*stime;
     for(size=0;size<NUMPNTS;size++)
          xxx += grazed(dubsize[size],stime)*sizedist[size];
     return (xxx/stime);
}

double    grazed(double n0, double stime)
{
     return( (BETA*KKKK/ALPHA) *
          ( log( ((ALPHA*n0+KKKK*(BETA-ALPHA))
               *exp((BETA-ALPHA)*stime)
               - ALPHA*n0) / (KKKK*(BETA-ALPHA)) ) )
               - BETA*(BETA-ALPHA)*stime*KKKK/ALPHA );
}
```

7.4.2 Code Details

Line 1: Two arrays, `sizes1[]` and `sizes2[]`, are used to hold the size distribution. First at line 1.a I initialize two pointers, declared within `main()`, to point to these arrays. One of these is passed to the function `InitSizeDist()` for initialization. Then within the time loop the size array is repeatedly updated at line 1.b, with `s1` being the array of old values, and `s2` being the array of new values. The pointers are then switched at line 1.c so that the array that used to hold the old values can be given the next set of updated values. A simpler, but less pedagogically useful, alternative calls `UpdateSizeDist()` a second time with the roles of the two pointers `s1` and `s2` reversed.

Line 2: These three arrays contain quantities that will be used repeatedly throughout the run. Because the quantities are functions of patch resource size only, and do not change during the calculation, the quantities are calculated once and stored. This tabulation provides easy access and saves computer time by not having to perform redundant computations. `dubsize[]` holds the resource value for each size index, and `growth[]` and `graze[]` hold values for the functions $g_e(s)$ and $-g_o(s)$, respectively, for each size index.

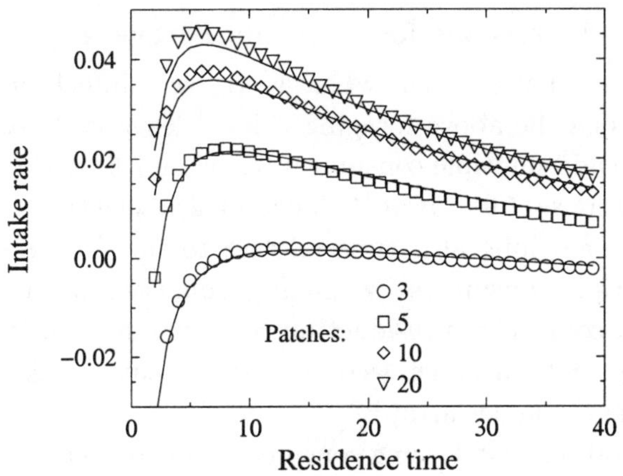

Figure 7.10. Results from the numerical integration of the size-structured model compared with the forager simulation's average intake rates. The size-structured model does a bit better than the simplistic model (compare with figure 7.7 on page 169).

7.4.3 Size-structured Model Results

I chose the value of `MAXTIME` such that convergence to the steady state patch distribution was assured. Although the size-structured model should describe the full range of the simulation's dynamical behavior, in the runs shown here I am not interested in the transient dynamics.

A comparison with the forager simulation results presented in figure 7.7 with the results of the size-structured model presented above is made in figure 7.10. Data points represent the simulation results and the solid curves are the size-structured model predictions. I used a small enough time and size step such that convergence to a stable solution was found (see exercise 7.9). Agreement between simulation results and the size-structured model results are somewhat improved over the first, crude model of section 7.1, especially for systems with few resource patches.

However, it is astonishing how well the simplistic model agrees with the simulation results compared to the agreement of the detailed size-structured model. One can argue that adding size-structured details to the analytic description has not helped a bit, and we should reach for the simplest model to address the question at hand. Of course this argument needs to be tempered in that the utility of the detailed model is likely contained in predicting variances (see exercise 7.11).

7.5 Extensions for Multiple Foragers

Let's look at what happens when we throw a bunch of consumers into the system. Using the above foraging rules, I leave it as an exercise to modify the code to add multiple consumers (exercise 7.5) within a spatially explicit habitat (exercise 7.6). Here I dangerously, scientifically speaking, change more than one thing at a time relative to the previous simulation. First, I add multiple consumers by making `consumer` an array `NCONS` elements long. Each consumer is resident on some patch (multiple patch occupancy is allowed), and consumer feeding and movement takes place sequentially through the consumer array.

Second, rather than forcing consumers to stay on a patch for a fixed length of time, I modify the movement rule such that consumers respond to the amount of resource on a patch. This rule seems completely reasonable: If a consumer finds a patch with high resource, then it should stay relatively longer than on a lower resource patch, leaving both patches when the resource is depleted to similar levels (Pyke 1983). Movement is now resource dependent; I add the following line of code after the consumer removes the resource from its patch in `Feed()`

```
if(con->patch->size<MIN_RESRCE) con->mvflag = 1;
```

`MIN_RESRCE` is some arbitrarily chosen resource level (below I use 0.2), and I also added a new member, `mvflag` (representing a movement flag), to the consumer's `con_def` structure. Each time step the consumer's movement flag is checked, and if set the consumer moves to a neighboring patch.

Results of two simulation runs with different consumer numbers, `NCONS`= 20 and `NCONS`= 40, are shown in figure 7.11. The habitat is a linear array of 200 patches, and each consumer is placed on a patch chosen randomly from the leftmost 100 patches. All consumers move to the right when leaving a consumed patch. What happens is that the consumers pile up on top of one another, sweeping through the habitat in consumer groups. Each group is stable (except when groups collide), in the sense that a consumer in front of the group would slow down because of the high resource density, and a consumer in the back of the group speeds up because of the low resource density. All grouping phenomena resulting from this model are mediated solely through individual consumer interactions with its resource, not through "social" consumer–consumer interactions. This collective dynamic arising from resource-dependent movement has been pursued mathematically for many years by people working on animal grouping models (e.g., Keller and Segel 1971, Gueron and Liron 1989).

One aspect I have not examined in this extension is the effect of grouping,

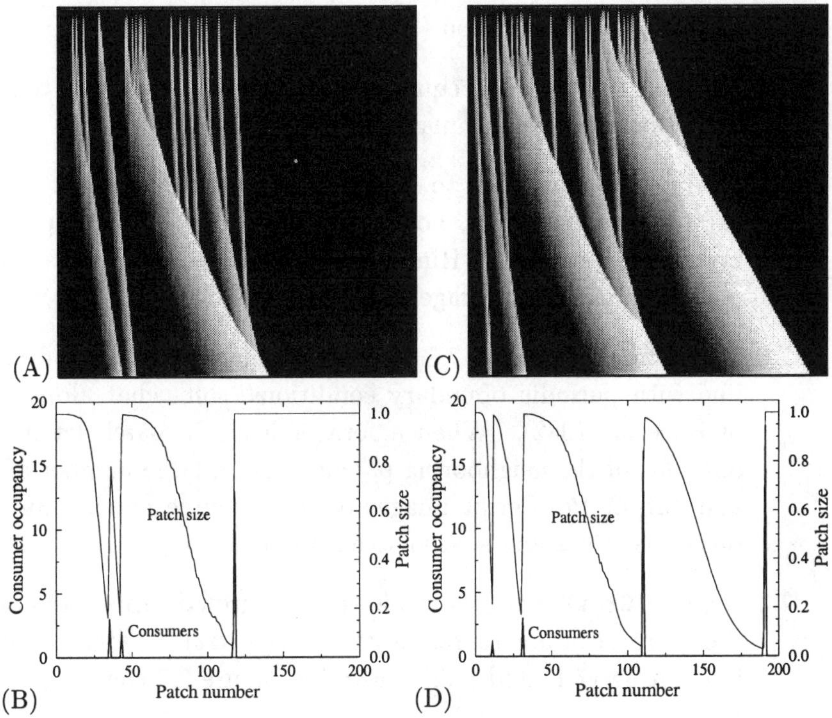

Figure 7.11. Grouping behavior resulting from resource-dependent movement rules. Simulations are run on one-dimensional habitats of 200 patches for 200 time steps with (A,B) 20 and (C,D) 40 consumers. Panels (A,C) are space–time images (increasing time downwards) of patch resource densities (dark pixels = high resource). Trails left behind by consumers converge, producing the grouped consumer distribution at the end of the simulations shown in panels (B,D).

or even the resource-dependent movement rule, on an individual consumer's intake rate. Although the full impact of consumer grouping on ecological processes remains to be fully worked out, initial investigations indicate that grouping can lead to the evolutionarily stable coexistence of foraging strategies (Wilson and Richards, in press).

7.6 Exercises

7.1 Compare the approximate model's estimate of r^* from equation (7.3) with measurements from the simulation. Does the approximate model do a good job?

7.2 Derive equation (7.8).

7.3 Derive $g'(T)$ in equation (7.11).

7.4 Examine $R''(T)$ when equation (7.10) holds. Can you tell analytically that it is really a maximum? (See section 7.2).

7.5 Modify the simulation to incorporate multiple foragers. As a function of residence time, how many foragers can be supported on a system of N patches? (Hint: One simple assumption regarding "support" is a positive average intake rate.)

7.6 Expanding on exercise 7.5, connect the patches in a one-dimensional line with periodic boundary conditions, somewhat along the lines of Kareiva (1982). When a forager leaves a patch let it move to only one of the neighboring patches. Use a large number of patches and output PostScript images of the resource patches and forager positions. Does any spatial structure develop?

7.7 Replace `GrowPatch()`'s crude logistic growth approximation with the more accurate representations $r(t + \Delta t) = f(r(t))$ defined by integrating (7.1) and (7.2). How does figure 7.7 change?

7.8 Make the forager simulation more efficient by updating unoccupied patch resource values only when the forager lands on a new patch.

7.9 In the size-structured numerical integration program, decrease by one-tenth (a) Δt, (b) Δs, and (c) Δt and Δs. How does each of these changes affect the results and numerical stability of the central differencing scheme? How about increases by 10 in each?

7.10 As outlined in section 7.5, modify the foraging simulation such that a forager leaves a patch when the patch resource level drops to some base resource level. For an isolated consumer measure the average intake rate as a function of the base resource level. Compare spatial (movement in one direction with periodic boundaries) and nonspatial (randomly chosen new patch) cases for a variety of habitat sizes. How does the optimal base resource level vary?

7.11 What is a good estimate of the patch size variance for the simple model of section 7.1? Measure and compare the mean and variance in patch size for the simple model, the simulation, and the size-structured patch model, as a function of the consumer's residence time. Examine cases $N = 3$, $N = 5$, and $N = 10$. Is the detail of the size-structured model important?

8

Maintenance of Gynodioecy

Programming components covered in this chapter:

- Runge–Kutta numerical integration
- character variables and strings
- command line arguments

Gynodioecy is a form of mating system in plants in which a single species has two types of individuals, those having only female flowers (called pistillate flowers) and those having hermaphroditic flowers (called perfect flowers). I will focus on a specific model in which the gene for male sterility (hence the gene for being female) is inherited through the cytoplasm, therefore passed down only through the maternal line (Frank 1989, Gouyon, Vichot, and VanDamme 1991). Presence of the cytoplasmic gene turns off growth of pollen-producing structures, rendering the individual male sterile. More complicated situations arise in real systems, for example, the nucleus fights back with a restorer gene that negates the cytoplasmic male-sterility gene and turns pollen production back on (Frank 1989). There are also systems in which male sterility is inherited via the nucleus (Pannell 1997) or both the cytoplasm and the nucleus (van Damme 1985). Whatever the details, if the females produce more seeds than the hermaphrodites (because they reallocate resources from male function), then females outcompete hermaphrodites. However, females remain dependent on hermaphrodites as the source of pollen to fertilize their ovules, whereas the hermaphrodites survive quite well without the females. If females outcompete hermaphrodites to the point of exclusion, females perish too.

This chapter examines a series of biologically simplified models for gynodioecy from a population dynamics perspective (e.g., Pannell 1997). Much work over the last several decades has uncovered mechanisms for female–hermaphrodite coexistence using genetic approaches (e.g., Lloyd 1975,

Charlesworth and Charlesworth 1978, McCauley and Taylor 1997). This chapter focuses on the role of spatial structure in promoting coexistence (McCauley and Taylor 1997, Pannell 1997). I first examine a nonspatial model that shows no possibility of an equilibrium containing both genotypes at nonzero densities – hermaphrodites persist on their own, but the introduction of a few females drives the system to extinction. Next, I translate this model into a spatial, individual-based simulation, compare deterministic nonspatial predictions with simulation results, and examine whether space can facilitate coexistence of the two genotypes.

8.1 Population Dynamics of Females and Hermaphrodites

8.1.1 Model Formulation

Consider a system of plants with two mating types, hermaphrodites and females. Hermaphroditism is the ancestral condition; define the population density of hermaphrodites as H. I assume, for the sake of the model, that hermaphrodites cannot self-pollinate, meaning that although a hermaphroditic plant produces both ovules and pollen, it must nonetheless import pollen from other individuals to successfully produce seeds.

Assume that females, with population density F, possess a cytoplasmically inherited male-sterility gene that turns off male function and renders the plant female. The plant then reallocates the resource previously devoted to male function into increased seed production. Inheritance in this system is such that females *always* produce female offspring, and hermaprodite mothers *always* produce hermaphrodite offspring. The pollen donor, which is always a hermaphodite, has no influence on the mating type of its offspring. Effectively, the cytoplasmic male-sterility gene parasitizes the male side of hermaphrodites, then outcompetes hermaprodites in terms of seed production. The drawback for the females is their need for pollen to fertilize their ovules, and so complete exclusion of hermaphrodites is a bad thing. This chapter explores whether long-term coexistence of so-defined hermaphrodites and females is possible.

The model assumes logistic growth with a carrying capacity set by physical locations to put down roots. I call each of these locations a site. Female and hermaphroditic plants produce θ_F and θ_H ovules, respectively. Each hermaphroditic plant produces an amount π of pollen which is dispersed uniformly throughout the habitat of N sites. By the very nature of the problem, it is clear that assumptions involving pollen–ovule interactions will be critical to coexistence. I assume a saturating curve for the dependence of per ovule fertilization on total pollen received

$$\frac{\pi H/N}{P_0 + \pi H/N},$$

where P_0 is the total amount of donor pollen required to fertilize one-half of a plant's ovules. Once ovules are fertilized, a parameter g describes seed survival and germination. Finally, the model assumes overlapping generations with a mating-type-independent mortality rate μ.

The above assumptions lead to the coupled differential equations

$$\frac{dH}{dt} = g\theta_H(1 - H - F)\frac{H^2}{H_0 + H} - \mu H \tag{8.1a}$$

$$\frac{dF}{dt} = g\theta_F(1 - H - F)\frac{HF}{H_0 + H} - \mu F, \tag{8.1b}$$

where $H_0 = NP_0/\pi$. This model assumes that H and F are measured in units of plants per site, thus the carrying capacity is 1, giving the factor $1 - H - F$. Time can be nondimensionalized by assuming $t = t't_0$, where t_0 carries the units of time which we are free to choose arbitrarily, and t' is a dimensionless number. Replacing $t = t't_0$ into equation (8.1) yields

$$\frac{dH}{dt'} = g\theta_H t_0(1 - H - F)\frac{H^2}{H_0 + H} - \mu t_0 H \tag{8.2a}$$

$$\frac{dF}{dt'} = g\theta_F t_0(1 - H - F)\frac{HF}{H_0 + H} - \mu t_0 F. \tag{8.2b}$$

I decree that we choose $t_0 = 1/\mu$, giving us

$$\frac{dH}{dt} = r_H(1 - H - F)\frac{H^2}{H_0 + H} - H \tag{8.3a}$$

$$\frac{dF}{dt} = r_F(1 - H - F)\frac{HF}{H_0 + H} - F, \tag{8.3b}$$

where $r_H = g\theta_H/\mu$, $r_F = g\theta_F/\mu$, and I have dropped the primes for aesthetics. Thus the model dynamics depends on only three parameters, r_H, r_F, and H_0.

8.1.2 Equilibrium Analysis

Again, the hermaphrodites do not need the females to survive. We can see this independence by setting $F = 0$, which gives us the female-free system

$$\frac{dH}{dt} = r_H(1 - H)\frac{H^2}{H_0 + H} - H. \tag{8.4}$$

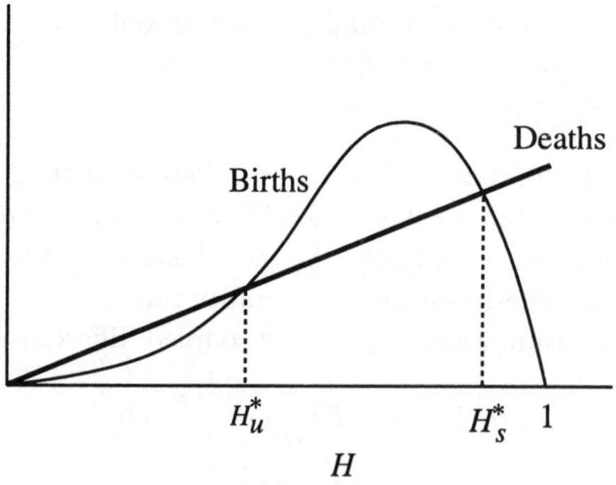

Figure 8.1. Schematic representation of the female-free model's bistability. Deaths occur with a density-independent mortality rate, but population growth at low densities is proportional to the square of population density (essentially the product of pollen density and ovule density). Hence, at very low densities ($H < H_u^*$) deaths exceed births and the net population growth rate is negative.

This equation has up to three biologically relevant equilibria, denoted by H^*, with either $H^* = 0$ or one of the nonzero densities

$$H^* = \frac{1}{2}\frac{r_H - 1}{r_H}\left[1 \pm \sqrt{1 - \frac{4r_H H_0}{(r_H - 1)^2}}\right]. \tag{8.5}$$

When $(r_H - 1)^2 = 4r_H H_0$ these two roots are identical, and for lower values of r_H the only real equilibrium is $H^* = 0$ (the others are imaginary). As $r_H \to \infty$ (seed production greatly exceeds plant mortality) the nonzero roots tend to 0 and 1.

When this model has three equilibria it exhibits *bistability*, or the presence of two stable equilibria. Figure 8.1 shows the birth and death functions that constitute equation (8.4). One of the key features for bistability is the H^2 factor in the birth term, called the Allee effect (Allee 1938; see Murray 1989); it hinders reproduction at low densities such that deaths exceed births for low population densities and the population approaches the stable equilibrium $H = 0$. Above the unstable equilibrium, H_u^*, births exceed deaths, and the population approaches the nonzero stable equilibrium, H_s^*.

Now consider the system of hermaphrodites and females. An equilibrium with both populations having nonzero densities H^* and F^* requires that the following two conditions (arising from setting the derivatives in equation

(8.3) to zero) hold simultaneously

$$1 = r_H(1 - H^* - F^*)\frac{H^*}{H_0 + H^*} \quad (8.6a)$$

$$1 = r_F(1 - H^* - F^*)\frac{H^*}{H_0 + H^*}. \quad (8.6b)$$

These conditions are identical except for the constants r_H and r_F, which means that, unless both constants are equal (an unlikely possibility since females can reallocate pollen production into greater seed production), both conditions cannot be satisfied simultaneously. Thus, the only possible equilibrium points have $F^* = 0$. Note that this result is independent of how ovule fertilization depends on pollen production as long as the functional form is identical for both females and hermaphrodites. For example, if we included the biologically reasonable self-pollination by hermaphrodites, then the functional forms would be different.

8.1.3 Runge–Kutta Numerical Integration

To explore the dynamics of the deterministic nonspatial gynodioecy model discussed above, we need a good, reliable numerical algorithm for integrating differential equations. The fourth-order Runge–Kutta method is probably the most common one; I introduce it first in general terms, then apply it to the gynodioecy model.

Suppose, for simplicity of discussion, we have a single population with density $n(t)$ whose dynamics are described by a single differential equation

$$\frac{dn}{dt} = f(n, t). \quad (8.7)$$

The function $f(n, t)$ describes the population's time rate of change. Suppose the dynamics defined by this differential equation are so complicated that an analytic solution cannot be obtained. What do we do? One option is a numerical solution, of which there are many approaches. Approaches I cover here stem from approximating the solution to (8.7) by a Taylor series expansion

$$n(t+h) = n(t) + h\left.\frac{dn}{dt}\right|_t + \frac{h^2}{2!}\left.\frac{d^2n}{dt^2}\right|_t + \frac{h^3}{3!}\left.\frac{d^3n}{dt^3}\right|_t + \frac{h^4}{4!}\left.\frac{d^4n}{dt^4}\right|_t + O(h^5), \quad (8.8)$$

taken out to four terms, an atypically large number for a Taylor series expansion. This expression yields the density at time $t+h$, given knowledge of

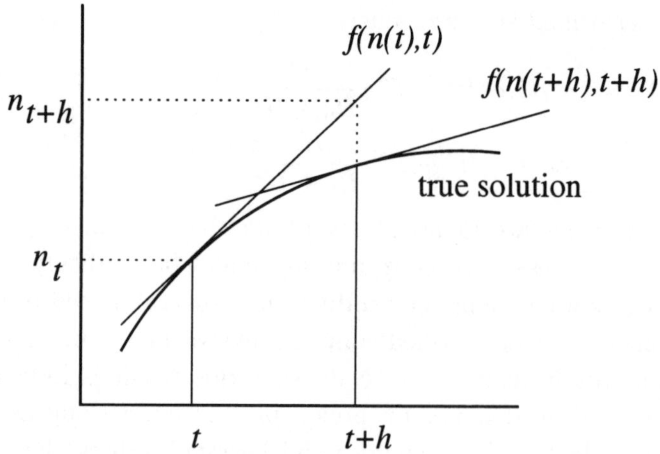

Figure 8.2. Schematic representation of Euler's numerical integration method. Calculation of n_{t+h} assumes the slope calculated at time t holds throughout the time interval t to $t+h$, which is clearly not the case. Runge–Kutta numerical integration (equation (8.13)) relaxes this assumption.

the function and its first few derivatives evaluated at the earlier time t. In the crude numerical approximations performed in earlier chapters I dropped everything past first order and approximated

$$n_{t+h} = n_t + hf(n_t, t). \tag{8.9}$$

Here I replaced the derivative with a function f and changed my notation from $n(t)$ to n_t to emphasize that estimates for $n(t)$ are obtained only at equally spaced intervals in time, n_t. This numerical scheme goes by the technical name of Euler's method. It's cheap; it's quick; it's dirty. One of Euler's method's main, good features is that it only requires evaluation of the known function $f(n_t, t)$ at the known time t to get an estimate for the value at the new time $t + h$, as in figure 8.2. It converges, which means if the step size h is halved, the error of the approximation is *guaranteed* to be at least halved. Convergence is a good thing. The method is also stable to some minimum degree (Atkinson 1978), meaning if it starts near the solution and is perturbed a tiny bit, then it will stay near the solution. In other words, errors are not badly amplified.

One downside to Euler's method is that the value used for the slope throughout the interval h is the value at the beginning of the interval leading to the value of n_{t+h} shown in figure 8.2. An improvement is to average the values at the beginning and the end of the interval, giving the expression

$$n_{t+h} = n_t + \frac{h}{2}\{f(n_t, t) + f(n_{t+h}, t+h)\}, \tag{8.10}$$

but this is a bit difficult to apply because n_{t+h} appears on both sides of the equation. Worse yet, one of the places it appears is inside a general function. An approximation that works is to use Euler's method to predict the value of n_{t+h} on the right side

$$n_{t+h} = n_t + \frac{h}{2}\{f(n_t, t) + f(n_t + hf(n_t, t), t+h)\}, \tag{8.11}$$

which looks exceedingly messy. Cleaning this expression up a little bit we can instead write

$$V_1 = hf(n_t, t), \tag{8.12a}$$
$$V_2 = hf(n_t + V_1, t+h), \tag{8.12b}$$
$$n_{t+h} = n_t + \frac{1}{2}V_1 + \frac{1}{2}V_2, \tag{8.12c}$$

which is a little more legible. Again, what happened was we replaced the estimate for the interval's slope (the slope at the beginning of the interval) with an estimate for the average slope throughout the interval by approximating the slope at the end of the interval as V_2/h. We hope that this estimate works better.

That's the concept. An extremely useful numerical integration scheme, called the fourth-order Runge–Kutta method, uses more complicated estimates for the slopes at different spots throughout the interval

$$V_1 = hf(n_t, t), \tag{8.13a}$$
$$V_2 = hf(n_t + \frac{1}{2}V_1, t + \frac{1}{2}h), \tag{8.13b}$$
$$V_3 = hf(n_t + \frac{1}{2}V_2, t + \frac{1}{2}h), \tag{8.13c}$$
$$V_4 = hf(n_t + V_3, t+h), \tag{8.13d}$$
$$n_{t+h} = n_t + \frac{1}{6}V_1 + \frac{1}{3}V_2 + \frac{1}{3}V_3 + \frac{1}{6}V_4. \tag{8.13e}$$

In the sense of the Taylor series expansion, the scheme is accurate through the fourth order, which makes for an amazing numerical scheme using only evaluations of the function defining the system dynamics. In other words, we never have to evaluate derivatives either analytically or numerically (a good thing) to obtain an extremely accurate numerical integration.

In the program presented below I extended the Runge–Kutta method to an arbitrary number of state variables, one differential equation for each, but the specific example I use is the gynodioecy model (equation (8.3)).

8.1.4 Code

```
#include <stdlib.h>
#include <stdio.h>
#include <math.h>

                  /* general numerical integration parameters */
#define NUMVARS    2
#define MAXTIME    40.0
#define OUTTIMEINC 0.1
#define STEPSIZE   0.01

                 /* specific MODEL parameters */
#define INITHERM   0.2
#define INITFEM    0.01
#define RH         40.0
#define RF         50.0
#define H0         3.5

double  state1[NUMVARS], state2[NUMVARS], stepsize = STEPSIZE;

int main(void)
{
    int         i, flag=1, cyclecnter=0;
    double      time=0.0, printtime=0.0,
                *oldstate, *newstate, *temp;

    oldstate = state1; newstate = state2;
    newstate[0] = INITHERM; newstate[1] = INITFEM;

    while(time<=MAXTIME)
    {           /* switch pointers */
        temp = oldstate; oldstate = newstate; newstate = temp;

        OneStep(time, oldstate, newstate, stepsize);

        time += stepsize;
        if(time>printtime)
        {    /* printout every OUTTIMEINC time units */
            printtime += OUTTIMEINC;
            printf(" %3.6f  %3.6f  %3.6f\n",
                  time, newstate[0], newstate[1]);
        }
    }
    return (0);
```

```
}

void OneStep(double time, double *in, double *out, double step)
{
    int         i;
    double      v1[NUMVARS], v2[NUMVARS], v3[NUMVARS],
                v4[NUMVARS], vals[NUMVARS];

    for(i=0;i<NUMVARS;i++) vals[i] = in[i];
    functions(time, vals, v1, step);

    for(i=0;i<NUMVARS;i++) vals[i] = in[i] + v1[i]/2.0;
    functions(time+step/2.0, vals, v2, step);

    for(i=0;i<NUMVARS;i++) vals[i] = in[i] + v2[i]/2.0;
    functions(time+step/2.0, vals, v3, step);

    for(i=0;i<NUMVARS;i++) vals[i] = in[i] + v3[i];
    functions(time+step, vals, v4, step);

    for(i=0;i<NUMVARS;i++)
        out[i] = in[i] + (v1[i] + 2.0*v2[i]
             + 2.0*v3[i] + v4[i])/6.0;
}

void functions(double time, double *state,
            double *derivs, double step)
{
    int         i;
    double      hh, ff;

    hh = state[0]; ff = state[1];
    derivs[0] = RH*(1.0-hh-ff)*hh*hh/(H0+hh) - hh;
    derivs[1] = RF*(1.0-hh-ff)*ff*hh/(H0+hh) - ff;
    for(i=0;i<NUMVARS;i++) derivs[i] *= step;
}
```

There will be no discussion of the code as there are no new computational issues other than the Runge–Kutta algorithm itself.

8.1.5 Nonspatial Model Results

Using the parameters defined in the code for the nondimensionalized model of equation (8.3), I ran the Runge–Kutta code for an initial hermaphrodite density of 0.2 per site and two initial female densities, 0.0 and 0.025. Figure 8.3 shows the results in two different formats. In panel (A) I plot the hermaphrodite and female densities as a function of time. Without females

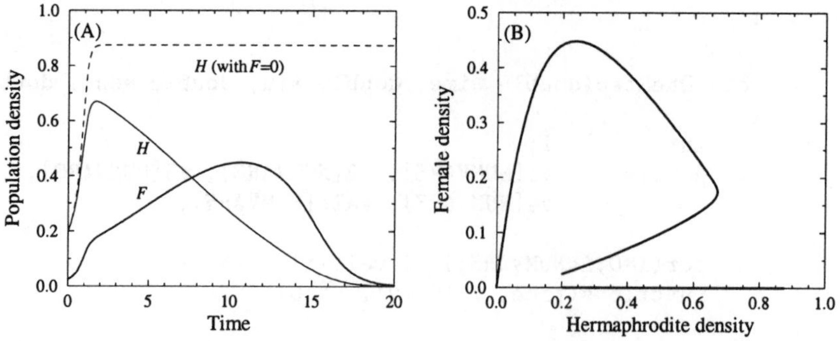

Figure 8.3. Numerical integration results for the nonspatial gynodioecy model (equation (8.3)) using $r_H = 40$, $r_F = 50$, and $H_0 = 3.5$. (A) Temporal dynamics of the female, F, and hermaphrodite, H, populations. Any nonzero initial density of females leads to the collapse of the system. (B) Phase plot of the female and hermaphrodite population dynamics.

the model demonstrates bistability, with both $H = 0.0$ and $H = 0.875$ being stable equilibria dependent on initial density H_0. However, with an initial nonzero female density all initial conditions have temporal dynamics that end up in the extinct equilibrium. Panel (B) plots the two dynamical variables, $H(t)$ and $F(t)$, against one another in a phase plot rather than a function of time. Here, ×s mark the initial locations in the phase plane for the two runs, and the dots identify the two stable equilibria. The nonzero equilibrium for H is only stable as long as there are no females, but even the smallest density of females takes the system on a long trajectory to the extinct state. This type of equilibrium is called a saddle point: Along one line the trajectory approaches the equilibrium (directly above the horse's backbone) and along another line the trajectory diverges from the equilibrium (perpendicular to the horse's backbone).

In the next sections I present a two-dimensional spatial simulation of the gynodioecy model and compare the deterministic results from the numerical integration with simulation results.

8.2 Pollen and Seed Dispersal in Space

One concise (although potentially incomplete) way of describing simulation rules is through a mathematical formulation. In the first part of this section I present spatial extensions of the nonspatial model, equation (8.1), falling roughly within the classification of reaction-diffusion models examined in detail in chapter 9. Turing's (1952) work on "morphogenesis" is an important foundation paper, followed closely by Segel and Jackson's (1972) ecological discussion. Given spatial extensions, I show the simulation code, then perform runs under a spatially mixed limit that makes contact with the nonspatial model predictions. I then examine simulation runs with spatial extensions.

8.2.1 Analytic Model Formulation

In this section the analytic model serves primarily as a concise statement of the simulation rules. No analysis of the analytic spatial model will be performed (although see exercise 8.9).

When plants are distributed spatially, pollen is dispersed only a short distance rather than mixed panmictically (throughout the population) as assumed in the nonspatial model. Local variation in hermaphrodite density, and hence pollen production, leads to a spatially varying fertilization probability. An important measure for both hermaphrodite and female populations is the pollen imported to location x from all other locations

$$P(x,t) = \pi \int_{\{space\}} K_p(y,x) H(y,t) dy. \tag{8.14}$$

$K_p(y,x)$ is called a *dispersal kernel* and represents the fraction of the pollen produced at location y, $\pi H(y,t)$, exported to location x. For example, if the habitat has a total area A and each plant's pollen is uniformly distributed over the habitat, then $K_p(y,x) = 1/A$ and $P(x,t)$ for all x becomes π times the average hermaprodite density, just as in the nonspatial model. Dispersal kernels are nice, conceptual functions representing the distribution of distances that organisms (or parts thereof) move in some given time interval. Although their use in theoretical ecology has been persistent (Skellam 1951, Gurney and Nisbet 1976, Levin and Segel 1985, Kot and Schaffer 1986, Othmer, Dunbar, and Alt 1988), they have recently become quite popular for representing organismal movement (Lewis 1994, Clark 1998, Wilson 1998) and given demonstrable empirical support (Clark *et al.* 1999).

Given the imported pollen $P(x,t)$, we then have the fertilization probability per ovule, $\phi(P(x,t))$, assumed identical for both genotypes. As above, we assume a functional form

$$\phi(P) = \frac{P}{P_0 + P}, \tag{8.15}$$

where P_0 represents the pollen required to achieve fertilization of one-half of the ovules. When pollen importation is low, each increment of pollen increases the fertilization probability by $1/P_0$, whereas at high pollen import the function saturates at 1.

With these definitions, the spatial model can be written as

$$\frac{\partial H}{\partial t} = r_H[1 - H - F] \int_{\{space\}} G_H(y,x)\phi(P(y,t))H(y,t)dy - H \tag{8.16a}$$

$$\frac{\partial F}{\partial t} = r_F[1 - H - F] \int_{\{space\}} G_F(y,x)\phi(P(y,t))F(y,t)dy - F, \tag{8.16b}$$

where the dispersal kernels $G_H(y,x)$ and $G_F(y,x)$ represent the probability that a seed produced at location y is dispersed to location x for hermaphrodites and females, respectively.

Approaches for analyzing these equations are limited. One approach we cover in chapter 9, but in a different context, is spatial linear stability analysis. This chapter will demonstrate that the predictions of nonspatial models do not hold, necessarily, for results of spatial models. In particular, a spatially explicit, individual-based simulation of the female–hermaphrodite system demonstrates coexistence of the two genotypes.

8.2.2 Simulation Model Formulation

The simulation model is a straightforward individual-based interpretation of the analytic model. A two-dimensional `DIMEN`×`DIMEN` grid of cells represents the spatial arena. Each cell contains at most one plant, which can be either a female or a hermaphrodite. The simulation's time step represents a single flowering period, during which all plants flower.

Pollen from hermaphrodites is dispersed uniformly over a circular region of radius `POLLDIST` centered about each plant (note exercise 8.10). Each ovule of both females and hermaphrodites are fertilized with an independent probability given by equation (8.15). All fertilized ovules successfully become seeds and are dispersed within a distance `HSEEDDIST` for hermaphrodites' seeds, and `FSEEDDIST` for females' seeds. Plants then die with probability `DEATHRATE`. Finally, one seed is chosen from all the seeds (if any) landing in unoccupied (or recently vacated) cells as the cell's new plant. All other seeds perish.

8.2.3 Code

```
     #include  <stdlib.h>
     #include  <stdio.h>
     #include  <sys/time.h>

     #define    MAXGENS         500
     #define    PRNTTIME    10
     #define    DIMEN        100
     #define    NPLANTS        (DIMEN*DIMEN)
     #define    PSOUTTIME    100
     #define    MIX        0
     #define    POLLPROD    (1.0)
     #define    POLLZERO    (3.5) /* used 5.0 for prior runs */
     #define    HSEEDS        (4)    /* used 20 */
     #define    FSEEDS        (5)     /* used 24 */
     #define    POLLDIST    (4)
     #define    HSEEDDIST    (2)
     #define    FSEEDDIST    (2)
     #define    DEATHRATE    (0.1) /* used 0.1 for prior runs */
     #define    INITDENSITY    (0.1)
     #define    INITHERMS    (0.8)

1.a) typedef struct plant_def
     {
         int    alive;
         int    herm;
         double inpollen;
         int    seeds;
         int    inseeds[2];  /*1-herms, 0-fems */
1.b) } Plant;
1.c) Plant plants[DIMEN][DIMEN];

     double deathrate;
     int    numh, numf, num_pnabes, num_hnabes, num_fnabes;
     int    polldist, hsdist, fsdist;
4.a) int    pnabes[4*POLLDIST*POLLDIST][2];    /* pollen neighbors */
     int    hnabes[4*HSEEDDIST*HSEEDDIST][2]; /* H seed nabes   */
     int    fnabes[4*FSEEDDIST*FSEEDDIST][2]; /* F seed nabes   */

     void ArgumentControl(int argc, char **argv);
     void InitStuff(void);
     void InitSeed(void);
     void InitPlants(Plant plnts[][DIMEN]);
     int  InitNabes(int *nabes, int dist);
     void CollectPollen(Plant plnts[DIMEN][DIMEN]);
     void DispersePollen(Plant plnts[DIMEN][DIMEN],
                   int idim, int jdim);
     int  NewInd(int ind);
```

```
    void SeedDisperse(Plant plnts[DIMEN][DIMEN]);
    void PutSeed(Plant plnts[DIMEN][DIMEN],
                 int *nabes, int num_nabes,
                 int idim, int jdim, int type);
    void KillPlants(Plant *plnt);
    void Germinate(Plant *plnt);
    void MixPlants(Plant *plnt1);
    void CountEm(Plant *plnt);
    void EPS_Header(FILE *, int, int);
    void EPS_Image(FILE *fileid, Plant plnts[][DIMEN]);
    void EPS_Trailer(FILE *);

5.a) int main(int argc, char **argv)
    {
        int   ttt;
2.a)    char  fname[20];
        FILE  *fopen(), *fileid;

        deathrate = DEATHRATE; polldist = POLLDIST;
        hsdist = HSEEDDIST; fsdist = FSEEDDIST;
5.b)    if (argc>1)  ArgumentControl(argc,argv);
        InitStuff();

        for(ttt=0;ttt<MAXGENS;ttt++) /* measurement period */
        {
            if(ttt%PRNTTIME==0)
            {
                CountEm(&plants[0][0]);
                printf("%5d H: %d   F: %d\n",ttt,numh,numf);
            }
            CollectPollen(plants);
            SeedDisperse(plants);
3.d)        KillPlants(&plants[0][0]);
            Germinate(&plants[0][0]);
    #if(MIX==1)
            MixPlants(plants);
    #endif
    #if(MIX==0)
            if(ttt%PSOUTTIME==0)
            {
2.b)            sprintf(fname,"xy%-1d.eps",ttt);
2.c)            fileid = fopen(fname,"w");
                EPS_Header(fileid,DIMEN,DIMEN);
                EPS_Image(fileid,plants);
                EPS_Trailer(fileid);
                fclose(fileid);
            }
    #endif
        }
    #if(MIX==0)
```

```
        fileid = fopen("xy_end","w");
        EPS_Header(fileid,DIMEN,DIMEN);
        EPS_Image(fileid,plants);
        EPS_Trailer(fileid);
    #endif
    return (0);
    }

     void ArgumentControl(int argc, char **argv)
     {
        int argcount, skip;

5.c)    for(argcount=1; argcount<argc; argcount+=skip)
        {
5.d)        if(argv[argcount][0] == '-')
            {
                skip = 1;
5.e)            switch(argv[argcount][1])
                {
5.f)            case 'm':
5.g)                sscanf(argv[argcount+1],"%lf",&deathrate);
                    skip++;
5.h)                break;
                case 'd':
                    if(argv[argcount][2]=='p')
                        sscanf(argv[argcount+1],"%d",&polldist);
                    else if(argv[argcount][2]=='h')
                        sscanf(argv[argcount+1],"%d",&hsdist);
                    else if(argv[argcount][2]=='f')
                        sscanf(argv[argcount+1],"%d",&fsdist);
                    else exit(0);
                    skip++;
                    break;
5.i)            default:
                    printf("**bad command line argument >%c< \n",
                             argv[argcount][1]);
                case 'h':
                    printf("Arguments (and default values):\n");
                    printf(" -h >> help\n");
                    printf(" -m %f >> death rate\n", deathrate);
                    printf(" -dp %d >> pollen disp\n", polldist);
                    printf(" -dh %d >> H seed disp\n", hsdist);
                    printf(" -df %d >> F seed disp\n", fsdist);
                    exit(0);
                }
            }
            else
            {
                printf("**invalid command line flag >%c<\n",
                       argv[argcount][0]);
```

```
                printf("try -h for help.\n");
                exit(0);
            }
        }
    }

    void InitStuff(void)
    {
        InitSeed();
3.a)    InitPlants(plants);
4.b)    num_pnabes = InitNabes(&pnabes[0][0],polldist);
        num_hnabes = InitNabes(&hnabes[0][0],hsdist);
        num_fnabes = InitNabes(&fnabes[0][0],fsdist);
    }

3.b) void InitPlants(Plant plnts[DIMEN][DIMEN])
    {
        int i,j;
        double xxx;

        for(i=0;i<DIMEN;i++)
        {
            for(j=0;j<DIMEN;j++)
            {
                plnts[i][j].alive = plnts[i][j].herm = 0;
    /*          if(i<DIMEN/2 && j<DIMEN/2)
    */          {
                    xxx = drand48();
                    if(xxx<(INITHERMS+INITFEMS))
                    {
                        plnts[i][j].alive = 1;
                        if(xxx<INITHERMS) plnts[i][j].herm = 1;
                    }
                }
            }
        }
    }

4.c) int InitNabes(int *nabes, int dist)
    {
        int i, j, tot=0, tester, ii;

        tester = dist*dist;
        for(i=-dist;i<=dist;i++) /* finds nabes within dist */
        {
            ii = i*i;
            for(j=-dist;j<=dist;j++)
            {
                if(ii+j*j<=tester && !(i==0 && j==0))
                {
```

```
                    nabes[tot++] = i;
                    nabes[tot++] = j;
                }
            }
        }
        return(tot/2);
    }

3.e) void CollectPollen(Plant plnts[DIMEN][DIMEN])
    {
        int iplant, i, j;
        Plant *plnt;

        plnt = &plnts[0][0];
        for(iplant=0;iplant<NPLANTS;iplant++,plnt++)
        {
            plnt->inpollen = 0.0;
            plnt->seeds = plnt->inseeds[0] = plnt->inseeds[1] = 0;
        }

        plnt = &plnts[0][0];
        for(i=0;i<DIMEN;i++)
            for(j=0;j<DIMEN;j++,plnt++)
                if(plnt->alive && plnt->herm)
                    DispersePollen(plnts,i,j);
    }

    void DispersePollen(Plant plnts[DIMEN][DIMEN],
                    int idim, int jdim)
    {
        int nn, i, j;
        double amnt;

        amnt = POLLPROD/(double)num_pnabes;
4.d)    for(nn=0;nn<num_pnabes;nn++)
        {
            i = NewInd(idim + pnabes[nn][0]);
            j = NewInd(jdim + pnabes[nn][1]);
            plnts[i][j].inpollen += amnt;
        }
    }

    int NewInd(int ind)
    {       /* periodic boundary conditions */
        if(ind<0) ind += DIMEN;
        if(ind>=DIMEN) ind -= DIMEN;
        return(ind);
    }

    void SeedDisperse(Plant plnts[DIMEN][DIMEN])
```

```
{
    int i, j, k;
    double fertprob;
    Plant *plnt;

    plnt = &plnts[0][0];
    for(i=0;i<DIMEN;i++)
    {
        for(j=0;j<DIMEN;j++,plnt++)
        {
            if(plnt->alive==1)
            {
                fertprob = plnt->inpollen/
               (POLLZERO+plnt->inpollen);
                if(plnt->herm==0)
                {
                    for(k=0;k<FSEEDS;k++)
                        if(drand48()<fertprob)
                            PutSeed(plnts,fnabes,
                    num_fnabes,i,j,0);
                }
                else
                {
                    for(k=0;k<HSEEDS;k++)
                        if(drand48()<fertprob)
                            PutSeed(plnts,hnabes,
                    num_hnabes,i,j,1);
                }
            }
        }
    }
}

void PutSeed(Plant plnts[][DIMEN], int *nabes, int num_nabes,
        int idim, int jdim, int type)
{
    int newspot, i, j;
                        /* uniform dispersal */
    newspot = 2 * (int)(num_nabes*drand48());
    i = NewInd(idim + nabes[newspot]);
    j = NewInd(jdim + nabes[newspot+1]);
    plnts[i][j].inseeds[type]++;
}
```

```
3.c) void KillPlants(Plant *plnt)
     {
         int i;

         for(i=0;i<NPLANTS;i++,plnt++)
             if(plnt->alive==1)
```

```
            if(drand48()<deathrate) plnt->alive = -1;
}

void Germinate(Plant *plnt)
{
    int iplant;
    double hseeds, fseeds, xxx;

    for(iplant=0;iplant<NPLANTS;iplant++,plnt++)
    {
        if(plnt->alive==0)
        {
            xxx = (double)(plnt->inseeds[0]+plnt->inseeds[1]);
            if(xxx>0.0)
            {
                plnt->alive = plnt->herm = 1;
                if(xxx*drand48()<(double)plnt->inseeds[0])
                    plnt->herm = 0;
            }
        }
        else if(plnt->alive==-1) plnt->alive = 0;
    }

}

void MixPlants(Plant *plnt1)
{
    int i, j, k;
    Plant *plnt;
    plnt = plnt1;

    for(i=0;i<NPLANTS;i++,plnt++)
    {
        j = NPLANTS*drand48();
        if(plnt1[j].alive)
        {
            plnt1[j].alive = plnt->alive;
            plnt->alive = 1;

            k = plnt1[j].herm;
            plnt1[j].herm = plnt->herm;
            plnt->herm = k;
        }
    }
}

void CountEm(Plant *plnt)
{
    int iplant;
```

```
        numh = numf = 0;
        for(iplant=0;iplant<NPLANTS;iplant++,plnt++)
        {
            if(plnt->alive)
            {
                if(plnt->herm)    numh++;
                else         numf++;
            }
        }

    }
```

8.2.4 Code Details

Line 1: Concurrent with defining a structure, it can also be elevated to the status of a new user-defined variable type. I define the structure used in this program as a new variable type by using the `typedef` declaration at line 1.a, calling this new variable type `Plant`. Once defined, I treat the new variable type identically to any other variable type. For example, at line 1.c I define a two-dimensional array of these variables.

Line 2: At line 2.a I declare an array of `char` variables used for holding values representing characters, and arrays of character variables for holding *strings* of ASCII characters. ASCII stands for "American Standard Code for Information Interchange" and represents the standard one-to-one mapping between printable (and nonprintable) characters and integer values. One byte (8 bits) is used for each `char` variable (a byte being the smallest unit of addressable memory).[1] Seven bits are used to store a character's numeric representation,[2] spanning the integer values 0 to 127. Thus, in keeping with ASCII, somewhere in my computer's disk is the one-computer-word hexadecimal series `6E657264` that translates to the word "nerd." I rarely use character variables in scientific programming, but I demonstrate one convenient application at line 2.b. This program outputs a sequence of spatial images as PostScript files, each of which needs a unique name. Here I give them the names `xy0.eps`, `xy100.eps`, `xy200.eps`, and so on, by

[1] Hard drives can store lots of text. My pocket dictionary has 57,000 entries, the average entry is about four lines, and the average line is about 40 characters. Since one byte holds one character, it would take a file roughly $57{,}000 \times 4 \times 40 = 9{,}120{,}000$ bytes $= 9.12 \times 10^6$ bytes $= 9.12$ megabytes to hold the entire dictionary. For scale, present computers ship with 10 Gbyte=10,000 megabyte hard drives.

[2] On Unix systems, type `man ascii` to see the codes; alternatively, most C reference manuals list them.

using the `sprintf()` function. This function is just like `printf()`, which outputs to the standard output, or `fprintf()`, which outputs to a file, except that `sprintf()` outputs to a string. The argument list includes as its first element the pointer to a character array (created by dropping the array's index) where the output will be placed. Inside the double-quotes is the control string `%-1d`, which means left-justify the integer in a field of a minimum of one space. I then use this string of characters as the filename opened at line 2.c.

Line 3: Notice that at line 1.c I declared the `plants` array as two-dimensional. Having passed the pointer to `plants` to `InitPlants()` at line 3.a, at line 3.b I declare the array as two-dimensional within `InitPlants()`. However, I often prefer to treat arrays as one-dimensional. Generally, multi-dimensional array references require a few more index calculations which can slow down simulations. For example, the `KillPlants()` function at line 3.c treats plants independently and I only need to increment the pointer to scroll through the entire array. In this case I pass the pointer to the first element of the array (see line 3.d). However some computations require explicit treatment of the two-dimensional configuration of plants (`CollectPollen()` at line 3.e is one example). In most cases switching between one- and two-dimensional representations is fine, but problems could arise if the array's memory is not allocated as a single contiguous block.

Line 4: In this program pollen and seeds are dispersed from a source cell to cells that sit within some distance away. Rather than repeatedly calculating which cells are within this distance, I declare three arrays to hold neighbor offset indices (relative to the present cell) at line 4.a. These arrays are initialized at 4.b by three successive calls to `InitNabes()` at line 4.c. Here I do the tedious calculation of offsets that represent cells within a circle of radius `dist` about the origin and store them in the neighbor arrays. Then I need only to scroll through this list of neighbors when calculating pollen import, for example, at line 4.d.

8.2.4.1 Command line input

Line 5: A particularly handy programming feature is passing arguments from the command line (typed within the Unix command window) to the program at execution time. Passing values for arguments avoids having to change and recompile the code whenever a parameter or two is changed. Line (5.a) defines the `main()` routine as having two arguments, one an integer variable and the other a pointer to a pointer to a `char` variable. On Unix machines, compilation of a program yields an executable binary file having the name designated after the `-o` flag (or `a.out` if no name is specified). For example, when compiling a program named `myprog.c` you might type

```
cc myprog.c -o xmyprog
```

and when running the program you type `xmyprog`. The compiler command `cc` itself is but one example of an executable binary, and the command line argument following the flag `-o` enables `cc` to read in a user-specified program file. Any C program can use this form of user–program communication to pass parameter values. Notice, however, that none of the previous programs have used this communication, yet compiled and ran, indicating that it can simply be ignored.

Suppose the above program was compiled and named `xgyno`, then executed using the line

```
xgyno -m 0.2 -dh 3
```

The Unix operating system passes to `main()` the number of arguments in `argc` (in this case five) and the arguments themselves via a pointer to an array of pointers, `argv`. Locations of these pointers and character values are shown schematically in figure 8.4. `argv` points to the first element of an array of pointers; each pointer in the array points to a string of characters holding the successive command line arguments. The operating system sets aside just enough space to hold each argument plus the *NULL* character (a special system-defined character represented by `\0` and having the ASCII representation 00) that terminates a string. The first argument passed is simply the name of the program that was executed, so if `argc` is one, then no arguments were passed on the command line.

If `argc` exceeds 1, then I pass both `argc` and `argv` to `ArgumentControl()` at line 5.b, an argument-processing subroutine.[3] Line 5.c loops over the command line arguments (skipping the name of the program). I always require a dash in front of the arguments passed to programs, so the first thing I look for is `-` at line 5.d; if no dash is present I warn the user and halt program execution. Upon finding a dash, control passes to the `switch` statement at line 5.e. If `-m` was entered, then control passes to line 5.f because the second character of the argument, `argv[argcount][1]` is an `m`. Line 5.g demonstrates reading the numeric value from the string of characters pointed to by `argv[argcount+1]` using the function `sscanf`. Upon reaching line 5.h, a `break` statement terminates the `switch` statement, program control falls back to the `for` loop, and on to processing additional arguments.

[3] `ArgumentControl()` was originally written many years ago for me by Wayne Groszko.

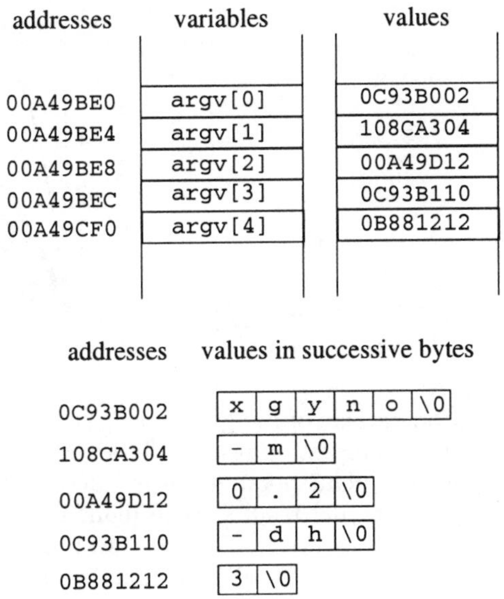

Figure 8.4. Pictorial representation of the arguments, **argc** and **argv**, passed to **main()** by the Unix operating system. **argc** indicates the number of arguments used in the line executing the program, and **argv** is a pointer to an array of pointers that access the command line arguments. This figure shows how the storage of the command line arguments is configured in memory.

If the command line flag was not one of the anticipated cases at line 5.e, control passes to the **default** case at line 5.i where I print out the available options and the parameters' default values. Without the **break** statement, execution proceeds with the lines in the next **case** statement.

Passing the name is neat in itself. Say you compile your code into the executable **xmyprog**. You can run this executable image and the operating system will pass the name **xmyprog** to the program through **argv[0]**. If you copy **xmyprog** to the file **xyourprog**, the operating system will pass the name **xyourprog** to the program. You can write a program to differentiate between these two different names, and thus perform different functions dependent on the name called to execute the program. However, I've never thought of a reason to use this feature in scientific programming.

8.2.5 Simulation Results

One way of benchmarking simulation programs written with spatial interactions is to remove the spatial correlations that build up within and between populations. Removing these correlations is rather easy: Pick up

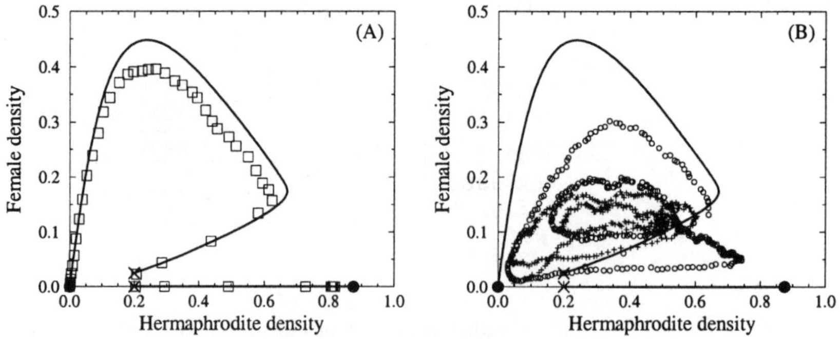

Figure 8.5. Comparison of the gynodioecy simulation results with the nonspatial numerical integration results (figure 8.3). (A) Homogeneous mixing of the simulation's plant population produces results qualitatively similar to the nonspatial model results: The plant population goes extinct. (B) When the populations are not mixed, the plant population persists with both females and hermaphrodites at nonzero densities.

each individual and assign it a new, randomly chosen location. I perform this scrambling (or homogeneous mixing) in `MixPlants()`, which is conditionally compiled by setting `MIX` to 1. Results are shown in panel (A) of figure 8.5, and provide overall good agreement with the deterministically expected trajectories. However, the location of the nonzero equilibrium is different from that obtained by numerical integration (see exercises 8.3–8.5).

Figure 8.5B shows the trajectories of the spatial simulation with two different initial conditions, both with $H = 0.2$ and $F = 0.02$, but one initially populating the entire lattice (○) and the other initially populating only a one-quarter section of the lattice (+) by uncommenting the `if()` statement in `InitPlants()`. Both runs demonstrate coexistence of the hermaphrodite and female populations.

Snapshots taken every 100 simulation steps are shown in figure 8.6. Both populations persist globally despite the fact that females always drive hermaphrodites extinct locally because the hermaphrodites always have an open front into which they can expand. Thus, the system is very spatially and temporally dynamic, but with an outcome that is very different from that predicted by the nonspatial models.

These results exemplify the interesting emergent spatial patterns found when simulating organismal interactions within a spatially explicit environment (Bascompte and Solé 1995, Rohani *et al.* 1997). Fascination with spatial patterns has driven much theoretical work over many decades (Zeigler

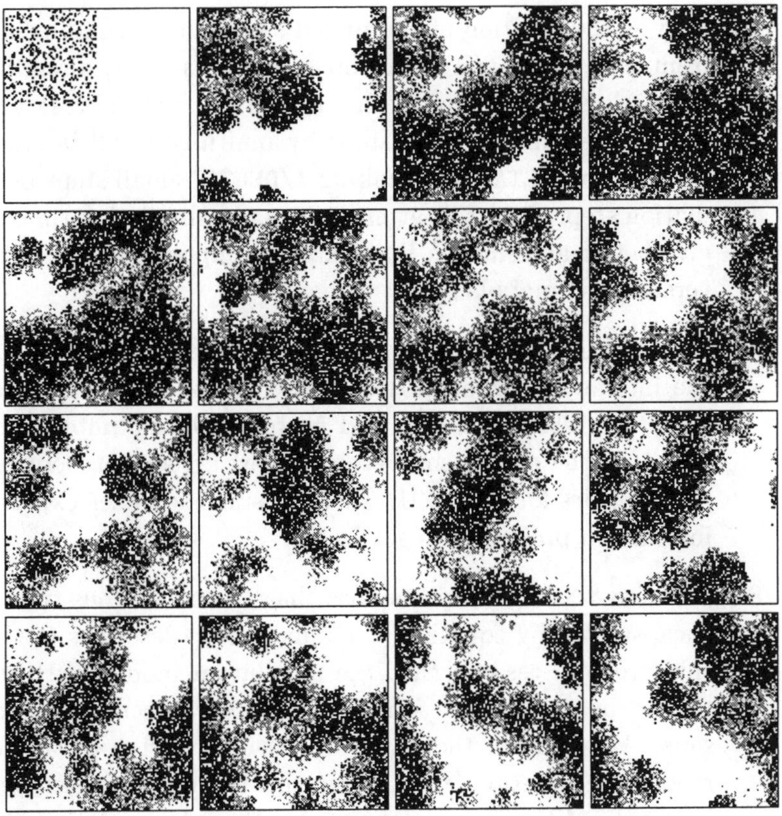

Figure 8.6. Lattice snapshots of the spatial simulation taken every 100 simulation steps (left to right, top to bottom). Dark pixels represent hermaphrodites, grey pixels represent females, and white pixels represent empty sites. On a local scale the dynamics are much like the nonspatial model (i.e., populations go extinct) but recolonization by neighboring populations allows population persistence on the global scale.

1977, Brown and Hansell 1987, de Roos, McCauley, and Wilson 1991, Durrett and Levin 1997), which we take up in the next chapter.

8.3 Exercises

8.1 Add a seed bank in the simulation by incorporating a seed survival rate. It would be easiest (in terms of CPU time) to implement the survival as a deterministic process. What is the effect on hermaphrodite–female coexistence?

8.2 Implement absorbing boundary conditions (pollen and seeds fall off the edges) and examine the influence of edge effects.

8.3 Why is the position of the mixed simulation's equilibrium different from that of the numerical integration? Is it the discrete time step in the simulation? Modify the simulation code to break a single simulation step into smaller steps by multiplying all interaction rates by a factor `DELTAT`, then taking `1/DELTAT` small steps per unit simulation step. Does this attempt to make simulation time continuous give better agreement? What are the implications for annual plants versus perennial plants?

8.4 As an alternative fix to the one suggested in exercise 8.3, replace the interaction probabilities $\mu\Delta t$ with $1 - \exp(-\mu\Delta t)$ (see Mangel and Tier 1993, Donalson and Nisbet 1999 for implementation details). Is the agreement better? Show agreement between the two probabilities for $\mu\Delta t \ll 1$. Justify using the latter expression when interaction probabilities get larger.

8.5 Exercise 8.3 implicitly assumes that the continuous-time model, as encapsulated by equation (8.1), accurately describes the biology. In this problem assume that the simulation model, with its discrete time steps, correctly describes the biological scenario of discrete seasons. Write down the "discrete-season" version of the female-free model (see equation (8.4)), determine the stable nonzero equilibrium hermaphrodite density, and compare the value with that observed for the mixed simulation model (figure 8.5A).

8.6 Implement the Runge–Kutta program to solve the continuous-time version of the food-chain model project (see project 6.9).

8.7 Incorporate a selfing rate by hermaphrodites (a hermaphrodite's pollen fertilizes its own ovules), letting selfed seeds have the same fitness as outcrossed seeds (i.e., no inbreeding depression). What is an appropriate fertilization function for hermaphrodites (see page 183)? For various selfing rates compare the numerical integration using this new function with results from homogeneous mixing results from an appropriately modified simulation model. How are the spatial dynamics affected by selfing?

8.8 Perhaps the effect of selfing is very important – here is a quick way to check if a detailed modification (exercise 8.7) might be interesting. `InitNabes()` disallowed selfing by excluding a hermaprodite from its own list of neighbors in the test

```
if(ii+j*j<=tester && !(i==0 && j==0))
```

Modify this line to allow selfing. Are the spatial results affected in any significant way?

8.9 Modify the simulation program to numerically integrate the analytic spatial model, much as we did for the analytic spatial competition model in section 5.3.2. Does coexistence of females and hermaphrodites result?

8.10 Try executing the simulation code with the line

```
xgyno -dp 10 -df 5
```

Does everything work? Are you sure? There is a bug in the program for too high of dispersal distances. Find and correct it (or them), then recompile and rerun the program.

9

Diffusion and Reactions

Programming components covered in this chapter:
• multiple header and program files • the Unix `make` utility

Interactions between organisms result in consumption and growth at an organismal level. Multiple interactions produce births and deaths yielding the dynamics of population-level measurements. Much of ecological theory translates the organismal interactions into population dynamical models with a few key assumptions, including the proviso that a large number of organisms interact within a *homogeneously mixed*, nonspatial environment. This set of assumptions, with roots in the field of chemical reaction kinetics, makes the mathematics easier: Analytically determining the dynamics of small populations having stochastic individinteractions and distributed over a spatially structured environment is not a trivial task!

Details of spatial structuring in ecological models are not always necessary, thereby justifying the homogeneously mixed assumption. Several cases exist demonstrating good, quantitative connections between mathematical models and well-controlled laboratory populations (e.g., whiteflies (Burnett 1958); blowflies (Gurney, Blythe, and Nisbet 1980, Stokes *et al.* 1988); Flour beetles (Costantino *et al.* 1995)). However, such connections often are predicated on making sure the laboratory populations satisfy the homogeneously mixed assumption of the models. In this way these important connections become much like the empirical tests of chemical reaction theory in which homogeneously mixed beakers of reacting chemicals are examined. If one stops mixing either of these systems of interacting entities, new and interesting phenomena may erupt.

There are but a few examples of empirical systems showing that spatial processes are operating within ecological systems. Huffaker's laboratory experiments on prey and predatory mites (Huffaker 1958) and Luckinbill's (1973) experiments on *Paramecium* and *Didinium* are classic examples of species interactions in spatial environments. An interesting recent spatial experiment was done by Holyoak and Lawler (1996) on *Didinium* and *Colpidium*. Probably the clearest examples of natural systems are invasions of exotic species into new habitat (Skellam 1951). However, Smith's (1983) analysis of census data on snowshoe hare populations demonstrate interesting natural spatial dynamics. The discussions in this chapter provide a background of the main concepts inherent to ecological interactions in a spatial environment, as opposed to an explicit account of any particular system.

Bringing together spatially explicit processes and ecological models of interacting organisms transforms a set of ordinary differential equations describing the temporal effects of species interactions into a set of partial differential equations also incorporating the spatial effects. An extremely valuable analytic tool used to approach these problems is spatial linear stability analysis,[1] which gives insight into the early temporal dynamics of spatially distributed interacting species.

Extending individual-based simulations into the spatial domain is rather easy, as demonstrated several times in earlier chapters. In addition to providing checks on the predictions of analytic models, their ability to produce the actual spatiotemporal dynamics of a model ecological system, given specific assumptions, enables developing insight into alternative model frameworks potentially more amenable to analytic methods.

In this chapter I connect simulation and analytic models using a range of increasingly complicated ecological models leading, in the end, to a spatial predator–prey model with diffusively moving individuals. In the first section I consider the conceptual and mathematical ideas behind diffusion, and obtain basic relationships of spatiotemporal processes from the diffusion equation. These relationships are examined by simulating a system of random walking organisms (examined briefly in chapter 4), one of the simplest interpretations of diffusive movement. Many introductory textbooks discussing spatial processes derive the analytic terms for diffusive movement from the assumptions of random walks (Turchin 1998). In the second section I extend diffusion concepts to situations where interactions between species also take place. What we discover is that the interplay between time scales of diffusion and the time scales of interactions leads to phenomena not at all governed by the predictions of nonspatial models.

[1] Part of the value of spatial linear stability analysis is that most everything else is analytically intractable.

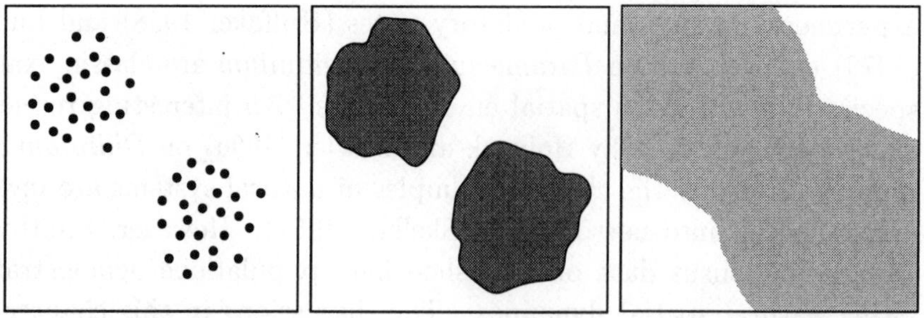

Figure 9.1. Dissolution of spatial pattern. Drops of dye sprinkled in water diffuse together over time, smoothing out small-scale spatial heterogeneities quickly, but taking longer times for larger heterogeneities. We simulate this process in section 9.1.

Simulations introduced in this chapter demonstrate another important component of structuring one's programs – splitting one computer program into an arbitrary number of program files. Ideally, each program file contains functionally related code, and all of these program files are compiled and linked together to produce one executable program.

9.1 Diffusion and the Dissolution of Pattern

One of the most ubiquitous concepts arising from diffusion theory is that spatial patterns are dissolved through the homogenizing process of diffusion. Conceptually, imagine a drop of black dye dripped into a beaker of clear water. As time goes on, the dye spreads throughout the water, producing a beaker full of grey water. For a more interesting aspect of diffusion, now imagine two groups of drops of black dye in a large container of clear water (as in figure 9.1). As time goes on the small drops merge together and, gradually, the two groups merge yielding one container of grey water. Diffusion has this natural tendency to smooth out inhomogeneities, starting with those at the smallest spatial scale. Below I examine this dissolution of spatial pattern from both an analytic and simulation approach.

9.1.1 Analytic Model of Diffusion

The diffusion equation describes temporal evolution of a population distributed over a spatial habitat. For the moment, consider a substrate covered with a population of motile organisms small enough such that the population density can be described by a continuous variable $n(x,t)$, where x represents position in space and t is time. At a particular instant, the population density varies in space, and likewise, at a specific location in space, there is temporal variation in population density. Thus, population density changes with respect to two different dimensions – space and time. In the idealized case of noninteracting organisms moving diffusively on a spatial habitat, the diffusion equation relates the temporal variation in density to the spatial variation in density.

There are many derivations leading to the diffusion equation. My favorite derivation (from Murray 1989; see also Bartlett 1960) starts with a description of movement using a general dispersal kernel, $K(y, x)$, which is the probability that an individual leaving position y ends up at position x (Othmer, Dunbar, and Alt 1988; Wilson 1998). We already used a kernel to model the dispersal of pollen away from its source plant in chapter 8. Suppose that individuals pick up and move with a rate ϕ. This movement means that the loss of individuals from any particular location x is $\phi n(x, t)$, the product of the leaving rate times the local density. As well, an amount $\phi n(y, t)$ of the individuals at location y left there, and a proportion $K(y, x)$ landed at position x. Note that since all individuals must go somewhere, the integral of $K(y, x)$ over all final positions y must be one, or $\int K(y, x)dy = 1$. Then, for a one-dimensional system,[2] movement can be described by the dynamical equation for the population density

$$\frac{\partial n(x,t)}{\partial t} = \phi \int K(y,x)n(y,t)dy - \phi n(x,t). \tag{9.1}$$

The first term on the right-hand side represents individuals arriving at location x from all over, and the second term represents individuals leaving position x and going elsewhere. I really like dispersal kernels for modeling spatial movement: Kernels capture long-distance movement that happens at very fast time scales relative to all of the interaction time scales in an ecological model.

Many important ecological concepts have been built upon the diffusion model for movement, and these are the concepts I focus on in this chapter. Using the integrodifferential equation model for movement (9.1), we can recover the diffusion equation (e.g., Okubo 1980, Turchin 1998). We begin by performing a Taylor series expansion of $n(y, t)$ about position x

$$n(y,t) = n(x,t) + \frac{\partial n(x,t)}{\partial x}(y-x) + \frac{1}{2}\frac{\partial^2 n(x,t)}{\partial x^2}(y-x)^2 + \dots, \tag{9.2}$$

[2] This derivation is also applicable to higher spatial dimensions.

and throw it into equation (9.1)

$$\begin{aligned}\frac{\partial n(x,t)}{\partial t} &= \phi \int K(y,x) \left[n(x,t) + \frac{\partial n(x,t)}{\partial x}(y-x) \right. \\ &\quad \left. + \frac{1}{2}\frac{\partial^2 n(x,t)}{\partial x^2}(y-x)^2 \right] dy - \phi n(x,t) \\ &= \phi n(x,t) \int K(y,x)\, dy + \phi \frac{\partial n(x,t)}{\partial x} \int K(y,x)(y-x)\, dy \\ &\quad + \frac{\phi}{2}\frac{\partial^2 n(x,t)}{\partial x^2} \int K(y,x)(y-x)^2\, dy - \phi n(x,t). \end{aligned} \tag{9.3}$$

Now we have to think about what the terms in the last line mean. Of the terms on the right-hand side, the integral in the first term gives 1 since the kernel is normalized (because all individuals go somewhere) and the first term cancels out the last. The second term drops away if the modeled individuals have no directional preferences: The integral is the sum of positive and negative terms of equal magnitude. This condition requires, for example, that for any two locations equally distant from x, $y = x \pm \Delta x$, the kernel satisfies $K(x+\Delta x, x) = K(x - \Delta x, x)$. This condition is sometimes called a zero first moment for the kernel, and the kernel is then called symmetric. If the kernel is symmetric, then all that remains is the third term, giving the diffusion equation

$$\frac{\partial n}{\partial t} = D\frac{\partial^2 n}{\partial x^2}, \tag{9.4}$$

where I have dropped the explicit dependence on x and t. Parameter $D = \phi R^2/2$ is the diffusion coefficient, or diffusivity, where R is an organism's mean square displacement when it takes a step

$$R^2 = \int K(y,x)(y-x)^2\, dy, \tag{9.5}$$

assumed independent of location x.

There is a lot of literature on the diffusion equation, and its connection to the random walk process. Connecting diffusion and Brownian motion was one of Albert Einstein's three accomplishments that earned him nominations for a Nobel prize.[3] In the early 1800s, Robert Brown, a Scottish botanist, discovered diffusion phenomena by observing the movement of pollen grains suspended in water. Recent references to diffusion in ecological models, as well as derivations of the diffusion equation, include Okubo (1980), Murray

[3] Einstein won the 1921 prize "for his services to theoretical physics and especially for his discovery of the law of the photoelectric effect" (Pais 1982).

(1989), Banks (1994), and Turchin (1998). Mathematically rigorous derivations of results for diffusive processes given stepping rules include Kehr, Kutner, and Binder (1981) and Larralde *et al.* (1992).

In connection with diffusion smoothing out patterns, be they patterns of dye particles or independently moving organisms, notice that one solution to equation (9.4) is a constant uniform density, $n(x,t) = n^*$. Replacing this solution for the ns that occur on both sides of the equal sign gives the equality, $0 = 0$. This result is somewhat trivial, but it represents the ultimate distribution for organisms obeying a simple, diffusive process. A slightly more interesting question is: What are the temporal dynamics of small spatial perturbations about this equilibrium solution? The short answer is that these perturbations either grow or decay, meaning that the spatially uniform equilibrium is either unstable or stable, respectively, to the perturbations.

Formalizing the issue of spatial stability, we assume a spatiotemporal perturbation described by the function

$$n(x,t) = n^* + n_0 e^{-t/T} \sin \frac{2\pi x}{L}. \tag{9.6}$$

n_0 represents the amplitude of the initial perturbation at time $t = 0$. I assume $n_0 << n^*$, which says that the perturbation added to the uniform equilibrium is very small. The first factor, $e^{-t/T}$, is a time-dependent decay function, which multiplies the amplitude by $1/e$ when $t = T$, or by $1/e^2$ when $t = 2T$, and so on. The second factor is $\sin \frac{2\pi x}{L}$, which represents the spatial variation of the initial perturbation, meaning the population density varies sinusoidally in space between $n^* - n_0$ and $n^* + n_0$. Parameter L is the perturbation's wavelength and represents the peak-to-peak (and trough-to-trough) spatial distance. Figure 9.2 shows what these initial perturbations look like for short and long wavelengths with identical amplitudes.

Proceeding with the stability analysis and substituting equation (9.6) into equation (9.4)

$$\begin{aligned}
\frac{\partial}{\partial t}\left(n^* + n_0 e^{-t/T} \sin \frac{2\pi x}{L}\right) &= D\frac{\partial^2}{\partial x^2}\left(n^* + n_0 e^{-t/T} \sin \frac{2\pi x}{L}\right) \\
\left(-\frac{1}{T}\right) n_0 e^{-t/T} \sin \frac{2\pi x}{L} &= -D n_0 e^{-t/T} \left(\frac{2\pi}{L}\right)^2 \sin \frac{2\pi x}{L} \\
\frac{1}{T} &= \left(\frac{2\pi}{L}\right)^2 D \\
T &= \frac{L^2}{4\pi^2 D}. \qquad (9.7)
\end{aligned}$$

This expression, called a dispersion relation, states an important concept

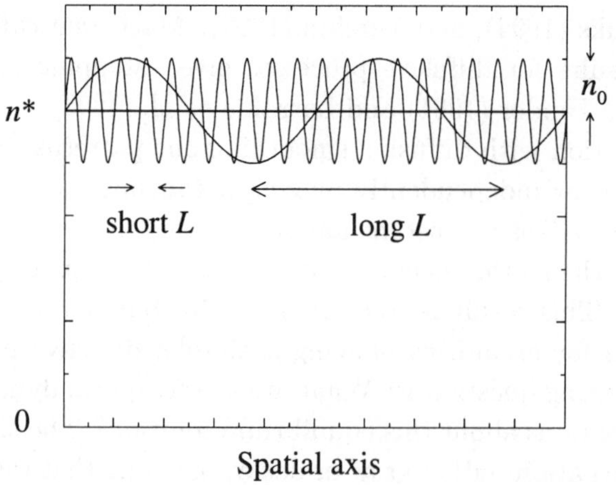

Figure 9.2. Initial spatial perturbations of wavelength L and amplitude n_0 about the uniform steady state n^*. Analysis of these small perturbations, represented by equation (9.6), leads to the relaxation time being $T \propto L^2$.

arising from the diffusion equation: The time taken for the spatial structure to dissipate depends on the *square* of the spatial structure's wavelength. Thus, doubling the size of a heterogeneity requires four times longer for it to diffuse away. Furthermore, the decay time is inversely proportional to the diffusivity D.

9.1.2 Simulation Model of Diffusing Organisms

The following program simulates a system of random walking particles on a two-dimensional square lattice of cells, each of which can contain at most a single particle. During one of the simulation's diffusive steps each particle on the lattice can potentially move to one of its four nearest neighbor cells. Hence, diffusion is represented here as a random walk (the basis of a common derivation of the diffusion equation (e.g., Turchin 1998)). Stepping rules shown in figure 9.3 are used to assure that all steps are performed in parallel, preventing any preferential or biased movement. A simple example of inadvertent biasing is choosing the particles' stepping order from one side of the lattice to the other side (see exercise 9.5). In the algorithm shown below, two situations result in a random walker being disallowed a step and returned to its originating site, depicted by the **X**s in the figure. First, if the walker chooses a neighbor that is occupied, it returns to its original site – even if the neighbor successfully moves to a different site. Second, if multiple

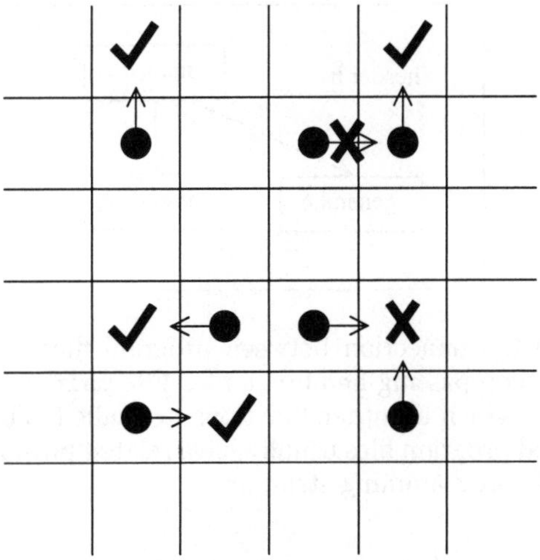

Figure 9.3. Simulation rules for random walking organisms. A move to a presently empty site is allowed unless another particle chooses it too.

walkers choose the same empty site, they all return to their original sites and the empty site remains empty. In part, I make these restrictions for code simplification, and, with a sacrifice in computational speed, one could design algorithms that work around them.

The model demonstrates the homogenizing influence diffusion has on spatial variation in population density using an initially heterogeneous distribution of individuals, then simulating the temporal evolution of the population's distribution.

9.1.3 Code

9.1.3.1 Overview of Multiple Program Files

The potential to structure C programs is high. A compound statement consists of related statements, a function consists of related compound statements, and a calling function consists of related function calls. Structuring extends to the level of program files.

Why break code into separate program files? First, the separate program files provide a better organization of your code into logical units. Once you produce a 3000-line program, it can be very difficult to find a specific 12-line routine buried in the code at line 1297. Second, using the `make` utility (discussed on page 223) allows you to compile only the program files that

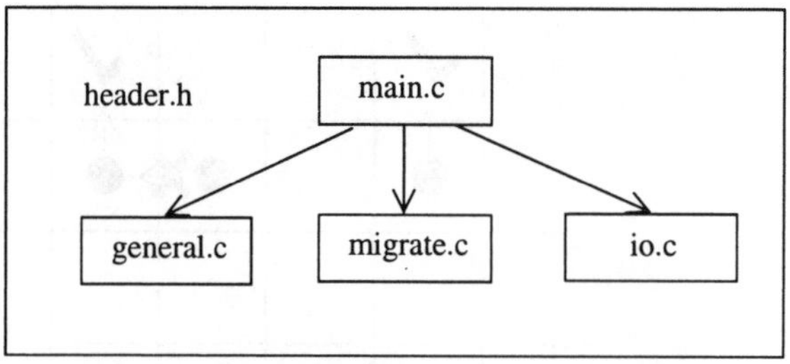

Figure 9.4. Connection between program files. All program files include the all-encompassing `header.h` file. File `main.c` contains the `main()` routine from which all other functions are called. Functions contained in the individual program files usually have related purposes, enabling yet another level of C programming structure.

have recent changes. If all you did is change an `i` to a `j` in this 12-line routine, you don't really need to compile the other 2988 lines of code.[4]

Just as a novelist doesn't *need* to break a book into chapters, a programmer doesn't *need* to break code into program files. However, the structure of both a book and a computer program might have logical breaks that form into natural units. A scientist has to understand what his or her code is doing and a clear understanding depends on clear code. Arguably, cause-and-effect are reversed – clear code is the product of a clear understanding of the process you want to program. Either way, organization into program files can help clarify this understanding.

Organization of the program files is illustrated in figure 9.4. There are two types of files that humans edit, the `.c` files and the `.h` file. A `header.h` file includes definitions for all global variables, function names, and structures, and included by all of the other program files. All of the `.c` files contain C code, with `main.c` containing the `main()` function. Dependencies of the code contained in the other files is shown by the directions of the arrows, for example, `main.c` calls functions from all the other program files.

One construct for building up these program files is to think of the `main.c` file as the trunk of a tree with a few branches coming out of it. In this case the branches are calls to subroutines. A good sort of tree has a few big branches and if you follow each of them, they in turn branch off into a few smaller branches. The `main.c` file represents only the first node where the trunk branches off. This analogy is useful for all subroutines – each subroutine represents a single node with multiple branches emanating from it.

[4] Minimizing compilation times is not very critical unless you have really, really big programs. Never have I had programs that took more than a few seconds to compile – usually the problem is how long the program takes to run.

9.1.3.2 The `header.h` *File*

The header file contains all the program lines that are common between all of the routines and program files constituting your code. These lines include all of your structure definitions, subroutine declarations, definition variables, and global variables. Additional header files could be defined holding features common to some program files, and thus included by those files, but not to other program files.

```
/*** HEADER.H ***/

#include <stdlib.h>
#include <stdio.h>
#include <math.h>
#include <time.h>
#include <string.h>

#define   NROWS      8
#define   NCOLS      640
#define   NSITES     (NROWS*NCOLS)

typedef struct siteinfo  /* SITE INFO STRUCTURE DEFINITION */
{
     int  prey;
     int  indirect;
     int  flag;
     struct siteinfo *nabes[4];
} Site;
Site habitat[NROWS][NCOLS];

          /*** MAIN.C ***/
void ArgumentControl(int argc, char **argv);

          /*** GENERAL.C ***/
void InitSeed(void);
void Initialize_Arena(Site *arena);
void Initial_Conditions(Site arena[NROWS][NCOLS]);
void TwoBands(int *array, int *spot, int size);

          /*** IO.C ***/
void EPS_Header(FILE *fileid, int width, int length);
void EPS_Row(Site arena[NROWS][NCOLS],
          int rownum, int skip, FILE *fileid);
void EPS_Trailer(FILE *fileid);

          /*** MIGRATE.C ***/
void DiffusePrey(Site *arena, double leftbias);
```

We previously covered most everything contained here, and the new functions defined by the function prototypes will be covered in discussions of individual program files. Only minor changes (not needing discussion) have been made in some functions for use in this specific program.

9.1.3.3 The `main.c` *File*

`main.c` contains the `main()` and `ArgumentControl()` routines. The latter routine is not reproduced here.

```
    /*** MAIN.C ***/
1)  #include "header.h"

    #define   TIMESTEPS (5000)
    #define   PREYDIFF  1
    #define   LEFTBIAS  (0.0)
    #define   TSKIP          10
    #define   XSKIP          1

2)  int       preydiff;
    double    leftbias;

    int main(int argc, char **argv)
    {
         int  i, step;
         FILE *imfileid;

         preydiff = PREYDIFF;
         leftbias = LEFTBIAS;
         if (argc>1)  ArgumentControl(argc,argv);

         InitSeed();
         Initialize_Arena(&habitat[0][0]);
         Initial_Conditions(habitat);
         imfileid = fopen("diffuse.ps","w");
         EPS_Header(imfileid,NCOLS/XSKIP,TIMESTEPS/TSKIP);

         for(step=0;step<TIMESTEPS;step++)
         {
              if(step%TSKIP==0)
                   EPS_Row(habitat,(int)(NROWS/2),XSKIP,imfileid);
              for(i=0;i<preydiff;i++)
                   DiffusePrey(&habitat[0][0], leftbias);
              if(step%500==0) printf(" time: %d\n",step);
         }
         EPS_Trailer(imfileid);
         return (0);
    }
```

Line 1: The header file is included at the top of each program file, in this case essentially inserting all of the lines of `header.h` directly into `main.c`.

Line 2: The variables `preydiff` and `leftbias` are defined globally with respect to all the functions defined in `main.c`. I define them this way to pass information between `main()` and `ArgumentControl()`. `preydiff` is used only locally, but the value of `leftbias` is passed to a function in another program file.

9.1.3.4 The `general.c` *File*

The program file `general.c` contains the routines associated with initializing the random number generators, the lattice structure, and the initial conditions for the population. I have not repeated the listing of `InitSeed()`.

```
     /*** GENERAL.C ***/

     #include "header.h"

     /*  NEAREST NEIGHBOR MAP.  If the array looks like:

                        (00) (01) (02) ....
                        (10) (11) (12) ....
                        (20) (21) (22) ....
                          .    .    .   .
                          .    .    .   .
                          .    .    .   .
                                       Top
     Then the neighbors are:  Left          Right
                                       Bot
     All sites have:
          Left: nabes[0]        Top:   nabes[1]
          Bot:  nabes[2]        Right: nabes[3]

     Sites that share a bond can be indexed as follows:
          [ site->nabes[i]->nabes[3-i] = site ]     */

1)   void Initialize_Arena(Site *arena)
     {
          int i, j;
          Site *origin;

          origin = arena;
          for(i=0;i<NROWS;i++)
          {
```

```
                for(j=0;j<NCOLS;j++)
                {
                     if(j>0) arena->nabes[0] = arena - 1;
                        else arena->nabes[0] = arena + NCOLS - 1;

                     if(i>0) arena->nabes[1] = arena - NCOLS;
                        else arena->nabes[1] = arena + NSITES-NCOLS;

                     if(i<(NROWS-1)) arena->nabes[2] = arena + NCOLS;
                        else         arena->nabes[2] = origin + j;

                     if(j<(NCOLS-1)) arena->nabes[3] = arena + 1;
                        else         arena->nabes[3] =
                                    arena - (NCOLS-1);

                     arena++;
                }
           }
      }
```

2.a)
```
      void Initial_Conditions(Site arena[NROWS][NCOLS])
      {
           int  i, j, k=0;
           int  mask[NCOLS];

           for(i=0;i<NCOLS;i++)
                mask[i] = 0;
           TwoBands(mask,&k,10);
           TwoBands(mask,&k,20);
           TwoBands(mask,&k,40);
           TwoBands(mask,&k,80);

           for(i=0;i<NROWS;i++)
           {
                for(j=0;j<NCOLS;j++)
                {
                     arena[i][j].prey = 0;
                     if(mask[j])
                          if(drand48()<0.5)
                               arena[i][j].prey = 1;
                     arena[i][j].flag = 0;
                }
           }
      }
```

2.b)
```
      void TwoBands(int *array, int *spot, int size)
      {         /* starting at *spot,
                     set up two bands of width size */
           int i;
```

```
    for(i=0;i<size;i++,(*spot)++)
         array[*spot] = 0;
    for(i=0;i<size;i++,(*spot)++)
         array[*spot] = 1;
    for(i=0;i<size;i++,(*spot)++)
         array[*spot] = 0;
    for(i=0;i<size;i++,(*spot)++)
         array[*spot] = 1;
}
```

Line 1: `Initialize_Arena()` is the most confusing routine of this entire program. It sets up the nearest-neighbor connections between the sites of a square lattice by placing site addresses into the **nabes** pointers. Within the inner loop, it sequentially considers the left, top, bottom, and right neighbors, and as implemented here, the lattice has periodic boundaries – the leftmost cells' left nearest neighbors are the rightmost cells. Likewise for the top, bottom, and right boundaries. Also demonstrated is the fast and free ability one has to do simple pointer arithmetic. The expression, `arena+1`, seemingly combines two very different variable types, a pointer to variable type `Site` and an integer. Your compiler might complain and give you a warning, but it will assume you meant to say, "I want the address of the next element in the `Site` array," and multiply the integer by the number of computer words that represents the size of a `Site` variable. This feature makes pointer arithmetic much easier, but potentially much more confusing than explicit array indexing.

Line 2: Routines `Initial_Conditions()` and `TwoBands()` set up the initial conditions for the lattice. `Initial_Conditions()` first zeroes a masking array used as the template for each row's initial configuration. It then adds onto this background a series of bands of various spatial scales using the subroutine `TwoBands()`.

9.1.3.5 The `migrate.c` *File*

The diffusion algorithm discussed here is the most important portion of the program. I used similar procedures in a variety of problems ranging from pure diffusive motion (Wilson and Laidlaw 1992) as well as the reaction–diffusion type of problems of population dynamics in spatial systems (Wilson, de Roos, and McCauley 1993).

```
     /*** MIGRATE.C ***/
     #include "header.h"

     void DiffusePrey(Site *arena, double leftbias)
     {
         int     i, dir;
         Site    *origin, *nnabe;
         double  xxx, bias[4];

     /* All sites have:  Left:   nabes[0]    Top:     nabes[1]
                         Bot:    nabes[2]    Right:   nabes[3] */
         bias[0] = 0.25 + leftbias; bias[1] = bias[2] = 0.25;
         bias[3] = 0.25 - leftbias;

         for(i=0,origin = arena;i<NSITES;i++,origin++)
         {
             if(origin->prey==1) /* consider move if present */
             {
3)               xxx = drand48()-bias[0]; /* choose step direction */
                 dir = 0; while(xxx>0.0) xxx -= bias[++dir];
                 nnabe = origin->nabes[dir];

1.a)                 if(nnabe->flag==0 && nnabe->prey==0)
                 {         /* empty and not tried before */
1.b)                     nnabe->indirect = 3 - dir;
1.c)                     nnabe->flag = 1;    /* flags step attempt */
1.d)                     origin->flag = 2;   /* individual moved */
                 }
1.e)                 else if(nnabe->flag==1) /* already stepped on */
                 {
1.f)                     nnabe->flag = 3;    /* flag mult steps */
1.g)                     nnabe->nabes[nnabe->indirect]->flag = 0;
                 }               /* reset prev stepper */
             }
         }
         origin = arena;
         for(i=0;i<NSITES;i++)
         {
2.a)             if(origin->flag>0)
             {
2.b)                 if(origin->flag==1) origin->prey = 1;
2.c)                 else if(origin->flag==2) origin->prey = 0;
                 origin->flag = 0;
             }
             origin++;
         }
     }
```

Line 1: Finally we see a use for the `flag` member of the `Site` structure. These flags are initially set to zero. At line 1.a, a step is only considered if the neighbor site is empty (no prey) and has not been stepped on previously. If these conditions are met, line 1.b sets the direction, `indirect`, from which the particle is stepping. Flags are then set at both cells indicating two states, `flag=1` to indicate a newly occupied cell (line 1.c), and `flag=2` to indicate a newly emptied site (line 1.d). One further case is considered at line 1.e – a step to a newly occupied cell. In this case the neighbor's flag is set to 3 at line 1.f, indicating that the cell was chosen multiple times, and the first particle's original cell's flag is reset to zero, at line 1.g, meaning that the particle remains there (and it will not be tested again).

Line 2: In 2.a I test for set flags. If a flag is set, I then process the conditions of the flag in lines 2.b and 2.c to correctly set the prey counts. If a flag is set to one, it means a particle ended up at that site, and if a flag is set to two, then a particle left that site.

Line 3: See exercise 9.6.

9.1.3.6 The `io.c` File

The letters "io" stand for input–output; in this program file I include the PostScript routines, `EPS_Header()`, `EPS_Row()`, and `EPS_Trailer()` presented in earlier chapters.

9.1.3.7 The `make` Utility

The `make` utility is a Unix-specific feature used to compile and link program files – compilers used under other operating systems, and even workshop packages for Unix, usually have some other way of merging the program files discussed in this section. Using `make` in ways more complicated than I show here is a skill better left in the hands of computer gurus. Their real benefit is cross-platform compilation, and understanding this use of makefiles can be very forbidding – many large books, into which I would not want to go, are devoted to describing the details.

I take a simple approach. Long ago, someone[5] showed me how to set up a basic `make` utility file, a text file on my computer which I always call `maker`, to compile code separated into different files into one executable program. Ever since then I have made no important modifications to my `make` file. I need nothing more than what I demonstrate in this chapter. There are small but important system-specific changes needed occasionally. These system

[5] Either Ken Gerke, Kim Wagstaff, or Paul Wellings (all at the University of Calgary).

dependencies also become important if you produce C code that others will compile and install on their own system and you want to minimize their effort by writing a system independent **make** file.

Below is my **maker** file for the diffusion program demonstrating the basic ingredients of how a program, structured into files of code, gets put together by **make**. None of my **maker** files are any more complicated than this one. In section 9.1.3.8 on page 225 I show how to compile with **make** and **maker**.

```
1)    OBJECTS=main.o io.o migrate.o general.o

2.a) bdiff: $(OBJECTS)
2.b)        cc -o bdiff $(OBJECTS) -lm

3.a) main.o: main.c header.h
3.b)        cc -c main.c

     io.o: io.c header.h
            cc -c io.c

     migrate.o: migrate.c header.h
            cc -c migrate.c

     general.o: general.c header.h
            cc -c general.c
```

Line 1: The first line defines a constant, `OBJECTS`, as the list of four `.o` files, which is replaced wherever it appears by its definition when the **make** file is executed, much like the defined variables used in C programs up to now.

Line 2: Line 2.a tells the **make** utility that the executable file that will be created and run by the user, **bdiff**, depends on all the files listed in `OBJECTS`. Line 2.b tells the utility how to build the executable (**bdiff**) using all of these `.o` files. Think of this line as very similar to the way we have compiled code up to this point.

Line 3: Line 3.a tells what files the object file `main.o` depends on – in this case, `main.c` file and the header file **header.h** called from within the `main.c` file (as seen below). Line 3.b instructs the **make** utility how to compile the program file. All the other program files are treated similarly.

9.1.3.8 Compiling the Program

Using the above **make** file called **maker**, the entire program is compiled with the Unix command[6]

```
make -f maker
```

Each program file will be compiled in turn, but compilation will stop immediately whenever fatal errors are found. Correct your mistakes in the program file containing the errors, then re-execute the **make** command. Repeat this process until you have successful compilation. If your compiler mentions errors regarding multiple declarations of globally defined variables, then you may need to declare all your global variables in only one program file and declare them as **extern**, meaning defined externally, in all the other program files.

9.1.4 Diffusion Results

Here I present PostScript images from the diffusion code. The spatial dimension is horizontal and temporal evolution is downward. As defined in **header.c**, the lattice is a thin, long strip with periodic boundary conditions (imagine a long soda straw joined at the ends) – the images are taken from one specific row of the lattice. Notice the initial bands of prey at the top of the figure. All structure at small spatial scales diffuses away very quickly leaving only the large structural features, in qualitative agreement with the analytic predictions of equation (9.7), as well as the conceptual expectations of figure 9.1.

9.2 Diffusion-Limited Reactions

When reactions are combined with diffusion, the two processes generate interesting spatiotemporal phenomena. At the heart of these phenomena are diffusion-limited reactions – interactions between organisms that are limited by spatial separation instead of biological processes. In other words, instead of reaction rates that are dependent on the densities of the interacting species (as is the case in nonspatial models), reaction rates become dependent on the diffusion rates that determine how quickly movement brings organisms together.

A straightforward extension of the one-species diffusion model discussed above includes a second species that interacts with the first. In the first

[6] A simple alternative to **make** that compiles the whole program is the line, `cc *.c -o bdiff -lm`, where `*.c` is a Unix command wild card representing all the relevant C files.

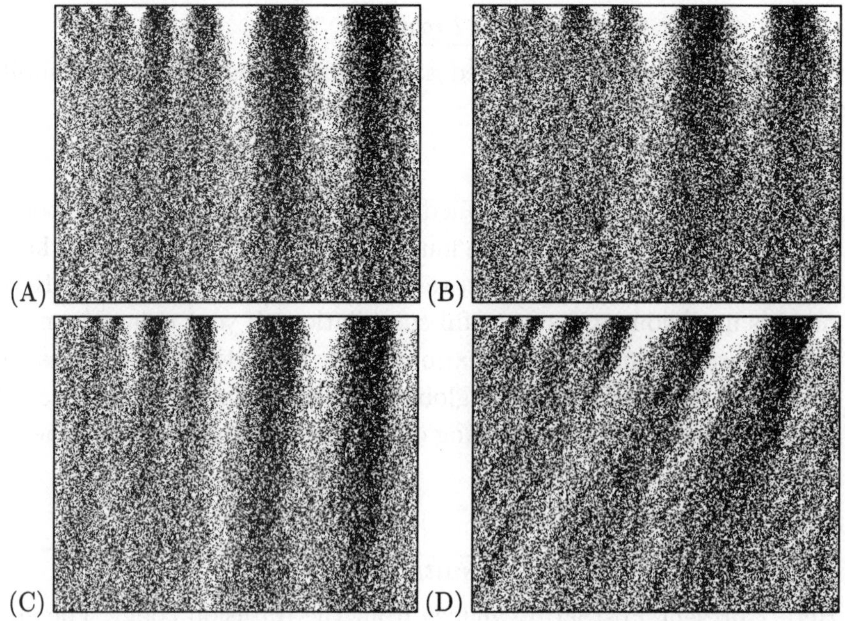

Figure 9.5. Diffusion of heterogeneities spanning various spatial scales. Panel (A) depicts diffusion of organisms with diffusivity $D = 1$ having an initial distribution generated by the `TwoBands()` initialization procedure. The run generating panel (B) used a doubled diffusivity. Note the faster dissolution of the spatial pattern. Panels (C) and (D) use values for `leftbias` of 0.01 and 0.05, respectively, for diffusivity $D = 1$. Patterns like these coming out of presumably symmetric diffusion algorithms are a sure sign of coding errors.

part of this section I model a simple *mutual annihilation* interaction: When two individuals of opposite species meet, both individuals disappear (e.g., Kopelman 1986). Examining this simplistic interaction clearly reveals interactions becoming limited by movement processes, and, although primarily pedagogical, this example forms the basis of several ecological interactions. For specificity I use the terminology of parasitism, with one species representing *unparasitized hosts* and the other representing *parasitoid eggs.* Interaction between the two produces the entity that is a parasitized host. Likewise, one species could represent unmated females and the other species unmated males; the interaction represents the forming of permanent mating pairs. Both examples may seem a bit biologically contrived, but in the latter part of this section I modify the interactions to produce the Lotka–Volterra prey–predator model.

Using this simple annihilation model, we'll find that, after an elapsed period of time, initial heterogeneities in species densities become increasingly evident. Excess individuals of one species or the other become increasingly separated from the other and interactions between the two species cease. While a biologically trivial interaction such as two-species annihilation in a spatial context seems pretty simple, analytically deriving macroscopic results from the microscopic rules is rather mathematically impenetrable (Doi 1976, Kang and Redner 1985, Clément, Sander, and Kopelman 1989).

9.2.1 Analytic Annihilation Models

9.2.1.1 Nonspatial Model

Call the two species H for host and P for parasitoid and assume the initial densities of the two species are identical. A nonspatial model of mutual annihilation would be written as the pair of ordinary differential equations

$$\frac{dH}{dt} = -\alpha HP \tag{9.8a}$$

$$\frac{dP}{dt} = -\alpha HP \tag{9.8b}$$

with initial conditions

$$H(t=0) = P(t=0) = N_0. \tag{9.9}$$

Equation (9.8) implies that the derivatives are equal to one another, and combined with identical initial values, we deduce that $H = P$ for all times. Thus we can replace one variable with the other, or

$$\frac{dH}{dt} = -\alpha H^2, \tag{9.10}$$

which can be integrated

$$\frac{1}{H(t)} - \frac{1}{N_0} = \alpha t \tag{9.11a}$$

$$H(t) = \frac{N_0}{1+\alpha N_0 t}. \tag{9.11b}$$

Thus, for times $t \gg 1/\alpha N_0$ the population decreases as t^{-1}.

9.2.1.2 Spatial Model

Now suppose that the two species are distributed over space. First, we could describe the dynamics of the hosts, $H(x,t)$, and parasitoids, $P(x,t)$, by a set of coupled partial differential equations

$$\frac{\partial H}{\partial t} = -\alpha HP + D\frac{\partial^2 H}{\partial x^2} \qquad (9.12a)$$

$$\frac{\partial P}{\partial t} = -\alpha HP + D\frac{\partial^2 P}{\partial x^2}. \qquad (9.12b)$$

This model represents one of the simplest possible reaction–diffusion models. Suppose we are given an initial sprinkling of hosts and parasitoids in space, at identical average densities, such that imperfect sprinkling produces some initial spatial variation. The solution to this set of partial differential equations is very difficult.

One way to see what happens to the system, at least qualitatively, is to change to new variables that describe the local average of the two species

$$\sigma = \frac{1}{2}(H + P) \qquad (9.13a)$$

$$\delta = \frac{1}{2}(H - P). \qquad (9.13b)$$

These new variables measure the local sums, $\sigma(x, t)$, and differences, $\delta(x, t)$, of the two species. Using these new variables we can transform the above set of partial differential equations into

$$\frac{\partial \delta}{\partial t} = D\frac{\partial^2 \delta}{\partial x^2}. \qquad (9.14a)$$

$$\frac{\partial \sigma}{\partial t} = -\alpha(\sigma^2 - \delta^2) + D\frac{\partial^2 \sigma}{\partial x^2}. \qquad (9.14b)$$

Imagine this scenario: A one-dimensional habitat has hosts sprinkled to the left of some location and parasitoids sprinkled on the right (figure 9.6A). If both species have equal initial densities in their respective regions, then, in terms of the transformed variables, σ is uniform across both regions but δ is positive in the host's region (where $H > 0$ and $P = 0$), and negative in the parasitoid's region (where $H = 0$ and $P > 0$; see figure 9.6B). Equation (9.14a) tells us that initial spatial variations within the two populations diffuse away just like the homogenization of spatial heterogeneities in the single-species (section 9.1). Interpretation of equation (9.14b) is more subtle. In regions where one species density is zero but the other is nonzero there clearly can be no interactions between the two species. Equation (9.14b) reflects this fact in that the reaction term $\sigma^2 - \delta^2$ is identically zero in these regions. The only place interactions occur, thereby decreasing the host–parasitoid sum, is at the interface between the species where $|\delta| < \sigma$ (the dotted line

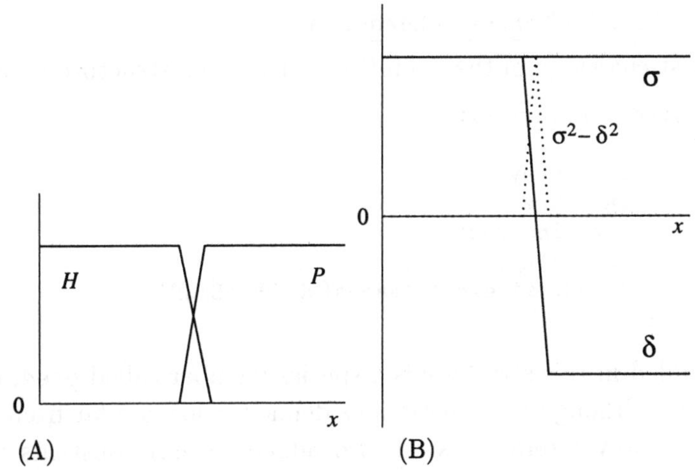

Figure 9.6. Panel (A) depicts an initial distribution having the two interacting species spatially separated. (B) In terms of the transformed variables, σ and δ (equation (9.13)), interactions occur only in the interfacial regions where $|\delta| < \sigma$ (the dotted line).

in figure 9.6B). In this sense the reactions become diffusion-limited because the interacting species occupy nonoverlapping spatial regions.

9.2.2 Simulation Code Modifications

Implementing the reactions of the reaction–diffusion simulation requires very few major changes to the diffusion code: Just add the other species variable to each site, move individuals of that species around the lattice, and interact individuals of the two species. Because the code is being built up from the previous version, I use the terms "prey" and "host," and "predator" and "parasitoid" interchangeably. The simplest way to extend the code is to copy the `DiffusePrey()` routine into a second routine, `DiffusePred()`, and make the necessary modifications. Assuming the prey and predator diffusivities are identical, it is then simple to modify `main()` to call `DiffusePred()`. A purist might not appreciate having two routines so similar in their code and would modify `DiffusePrey()` such that it could diffuse either species dependent on the value of a passed flag, but this change takes a little bit more work (see exercise 9.2). I also modified the output routines to produce two files, one for each species, and I now fill a pixel if any of the rows contains a member of the specific species.

9.2.2.1 Changes to header.h

The first change is in the definition of the **Site** structure variable

```
typedef struct siteinfo
{
	int  prey;
	int  pred;
	int  indirect;
	int  flag;
	struct siteinfo *nabes[NO_OF_NEIGH];
} Site;
```

All I added in this structure is a species member called **pred**, nominally for predator, although the interactions defined below are far from being typical predator–prey interactions. I also added several constants to **header.h**, some of which will be seen in the routines detailed below.

9.2.2.2 Changes to main.c

The next new addition to the reaction–diffusion code is a call in the **main()** routine to interact individuals of the two species. The call made in the time loop is

```
Interact(&habitat[0][0],&preycnt,&predcnt);
```

where **preycnt** and **predcnt** are variables local to **main()** that have the total species numbers passed back in them.

9.2.2.3 Changes to general.c.

The initial conditions for the two species differ considerably from those I used in the single-species diffusion code. In this case both populations are initialized with the same number of individuals uniformly distributed over the entire lattice. The following routine replaces the previous one.

```
void Initial_Conditions(Site arena[NROWS][NCOLS],
                        int *nprey, int *npred)
{
	int  i, j, total;
	Site *pnter;
	double    cosval, hprob, pprob;

	pnter = &arena[0][0];
	for(i=0;i<NSITES;i++,pnter++)
		pnter->prey = pnter->pred = pnter->flag = 0;

	if(itype==0)
	{
```

```
            total = NZERO*NSITES;
            (*nprey) = *npred = total;
            pnter = &arena[0][0];
            for(i=0;i<total;i++)
            {
                j = drand48()*NSITES;
1.a)            while(pnter[j].prey==1) j = drand48()*NSITES;
                pnter[j].prey = 1;

                j = drand48()*NSITES;
1.b)            while(pnter[j].pred==1) j = drand48()*NSITES;
                pnter[j].pred = 1;
            }
        }

        if(itype==1)
        {
            *nprey = *npred = 0;
            for(j=0;j<NCOLS;j++)
            {
                cosval = cos(2.0*M_PI*j/wavelength);
                hprob = NZERO*(1.0+AZERO*cosval);
                pprob = NZERO*(1.0-AZERO*cosval);
                for(i=0;i<NROWS;i++)
                {
                    if(drand48()<hprob)
                    {
                        arena[i][j].prey = 1;
                        (*nprey)++;
                    }
                    if(drand48()<pprob)
                    {
                        arena[i][j].pred = 1;
                        (*npred)++;
                    }
                }
            }
        }
    }
```

Line 1: These statements are the only ones that demonstrate anything new. The goal is to place one new individual onto the lattice for each pass through the `i` loop. An initial random choice for the placement site is made and placed into the variable `j`. The annotated lines check whether this site is occupied, if not the `while()` loop is skipped and the individual is placed. If the initial site is occupied then the `while()` loop is executed, a new choice for `j` is made, and the test in the `while()` loop is repeated. Only when the chosen site is unoccupied does execution drop from the loop.

Note that if we had set `total` larger than `NSITES` a problem would have arisen once all the sites became occupied: The condition in the `while()` loop would never be satisfiable, resulting in an infinite loop. An infinite loop is one that never allows its execution to be ended.

9.2.2.4 Changes to `migrate.c`.

The `Interact()` and `MixPred()` routines are located in the `migrate.c` program file.

```
        void Interact(Site *arena, int *nprey, int *npred)
        {
                int  i;
                Site *origin;

                origin = arena;
                (*nprey) = *npred = 0;
                for(i=0;i<NSITES;i++,origin++)
                {
1.a)                    if(origin->prey) (*nprey)++;
1.b)                    if(origin->pred) (*npred)++;
2)                      if(origin->prey==origin->pred)
                                origin->prey = origin->pred = 0;
                }
        }

void MixPred(Site *arena)
{
        int i, j, k;
        Site *origin;

        origin = arena;

        for(i=0;i<NSITES;i++,origin++)
        {
                j = NSITES*drand48();

                k = origin->pred;
                origin->pred = arena[j].pred;
                arena[j].pred = k;
        }
}
```

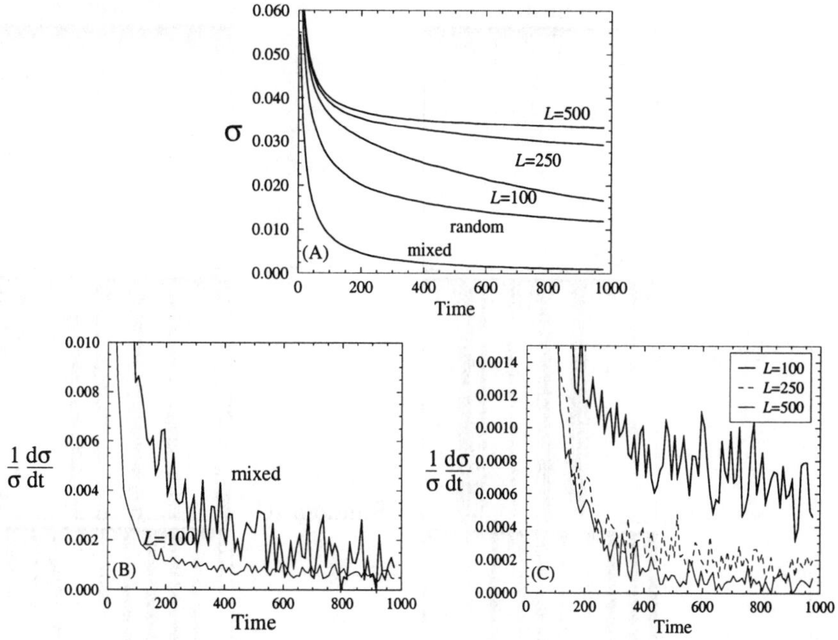

Figure 9.7. (A) Population counts and (B,C) measured interaction rates averaged over 20 simulation runs as a function of time. The population counts represent an integration over the interaction rates, hence the population curves are smoother. The message is that the reactions become diffusion-limited when the interacting organisms are separated spatially (i.e., the interactions cease).

Line 1: The purpose of these two lines is to take the census of prey and predators in the arena.

Line 2: This statement is the only interaction between the two species. The only time a change in state occurs is when the site is occupied by both prey and predator, in which case both are annihilated.

9.2.3 Diffusion-limited Reaction Results

Figure 9.7 presents the spatially averaged temporal dynamics of $\sigma(x,t)$, further averaged over 20 runs, for cases of spatial mixing, random initial conditions, and three different initial perturbation wavelengths (as depicted in figure 9.2). Mixing just the parasitoid (synonymously predator) species enhances interactions and causes a rapid decline in the host and parasitoid populations, as seen in figure 9.7A. High interaction rates are maintained for an extended period in the mixed case, demonstrated through the measured

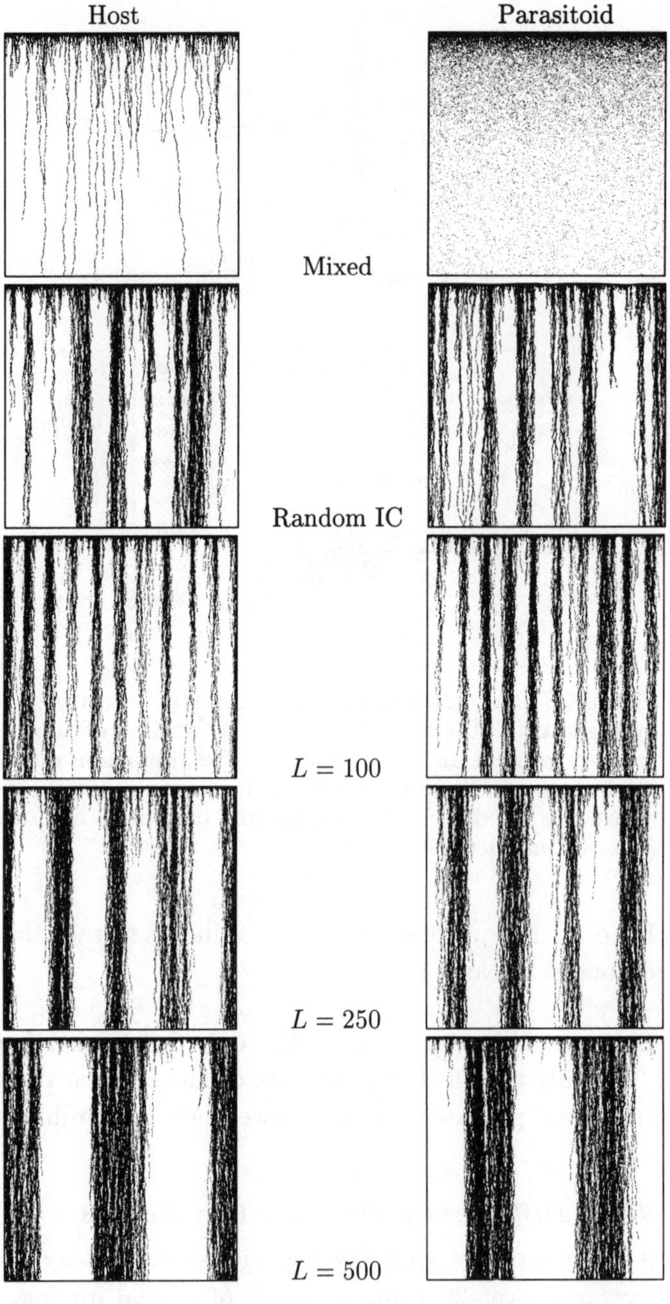

Figure 9.8. Host and Parasitoid species population distributions. Initial heterogeneities in prey and predator densities become amplified as annihilation interactions take place, and further interactions occur only at the species boundaries.

per capita interaction rates shown in figure 9.7B. At the other extreme, figure 9.7C indicates that species interactions quickly decrease to very low rates when the system is initialized with long wavelength perturbations, resulting in population numbers that remain high for an extended period of time (reflected by σ for $L = 500$ in figure 9.7A).

Spatiotemporal images in figure 9.8 from single runs vividly demonstrate the quantitative measurements of figure 9.8. Mixing the parasitoids causes a rapid decline in the populations, whereas all other cases result in the maintenance of the two populations for a very long time. This difference is due to the cessation of interactions via spatial segregation in the latter case, except at the edges of the bands where the populations are slowly eaten away.

9.3 Spatial Predator–Prey Model

9.3.1 Analytic Lotka–Volterra Models

9.3.1.1 Nonspatial Model Analysis

I now extend the simulation presented in the above sections to a predator–prey model, specifically the Lotka–Volterra model discussed in section 1.2. In a differential equation framework, the interactions are written as the pair of equations

$$\frac{dV}{dt} = \alpha V - \frac{\beta}{N} VP \tag{9.15a}$$

$$\frac{dP}{dt} = \epsilon \frac{\beta}{N} VP - \delta P, \tag{9.15b}$$

where V and P represent the prey (or victims) and predator population numbers, respectively. α is the per capita prey reproduction rate, ϵ is the conversion rate of prey into predators, and δ is the per capita death rate of predators. I use an aggregate parameter, β/N, for the attack rate per prey per predator. I envision V prey distributed over N predator search regions, hence V/N is the probability a predator finds a prey during a search. β is then the product of the predator's search rate and the probability of a successful attack upon each prey discovery.

The Lotka–Volterra model is the "fruit fly" of predator–prey theory: It is the simplest possible model that encapsulates predators attacking prey and converting them into predator offspring. An excellent introduction to more detailed predator–prey models was given by Stenseth (1980).

Setting the time derivatives in equation (9.15) to zero and solving yields the nonzero equilibrium densities $V^* = \delta N/\epsilon\beta$ and $P^* = \alpha N/\beta$. If the system is initialized close to this equilibrium, then the subsequent dynamics of the system define the equilibrium's stability. If, as time evolves, the system moves further away from the equilibrium, then the equilibrium is unstable to small perturbations. If the system moves towards the equilibrium, then the equilibrium is stable against small perturbations.

What is the stability of the Lotka–Volterra model's nonzero equilibrium? First set the initial densities equal to the equilibrium densities plus a small perturbation

$$V(t) = V^*(1 + v(t)) \tag{9.16a}$$
$$P(t) = P^*(1 + p(t)). \tag{9.16b}$$

Substituting these expressions into equations (9.15) yield

$$\frac{dv}{dt} = -\alpha p \tag{9.17a}$$
$$\frac{dp}{dt} = \delta v, \tag{9.17b}$$

assuming $v(t), p(t) \ll 1$ such that $v^2, p^2, vp \approx 0$. This assumption means that the perturbations are so small that any product of two perturbations is extremely tiny and can be ignored. In the case of the Lotka–Volterra predator–prey model we can proceed further by differentiating equation (9.17a)

$$\begin{aligned} \frac{d}{dt}\frac{dv}{dt} &= -\frac{d(\alpha p)}{dt} \\ \frac{d^2v}{dt^2} &= -\alpha\frac{dp}{dt} \\ \frac{d^2v}{dt^2} &= -\alpha\delta v, \end{aligned} \tag{9.18}$$

where equation (9.17b) was used in the last line. The solution to this second-order differential equation might be familiar; replacing into equation (9.18) the trial solution $v(t) = v_0\cos(\omega t)$ indicates that the trial solution works if $\omega = \sqrt{\alpha\delta}$. The implication of this periodic solution is that if we perturb the system from its equilibrium by adding a small amount of prey, $v_0 V^*$, the temporal dynamics of the system take the form

$$V(t) = V^*\left(1 + v_0\cos(\omega t)\right) \tag{9.19a}$$
$$P(t) = P^*\left(1 + v_0\sqrt{\frac{\delta}{\alpha}}\sin(\omega t)\right). \tag{9.19b}$$

We thus anticipate perpetual oscillations around the equilibrium with the amplitude of the oscillations being determined by the size of the initial perturbation.

An exact solution for the Lotka–Volterra model exists in the predator–prey state space. Dividing equation (9.15a) by (9.15b) yields the ordinary differential equation

$$\frac{dV}{dP} = \frac{\alpha V - \frac{\beta}{N}VP}{\epsilon\frac{\beta}{N}VP - \delta P} \tag{9.20a}$$

$$= \frac{V}{\epsilon\frac{\beta}{N}V - \delta}\frac{\alpha - \frac{\beta}{N}P}{P} \tag{9.20b}$$

$$\left(\epsilon\frac{\beta}{N}V - \delta\right)\frac{dV}{V} = \left(\alpha - \frac{\beta}{N}P\right)\frac{dP}{P} \tag{9.20c}$$

which can be integrated to give equation (1.3) from section 1.2, repeated here

$$\alpha N \ln\frac{P}{P_0} + \delta N \ln\frac{V}{V_0} = \epsilon\beta(V - V_0) + \beta(P - P_0), \tag{9.21}$$

where V_0 and P_0 are the initial prey and predator densities.

In figures 9.9A,B I show the population dynamics resulting from two numerical integrations using the Runge–Kutta algorithm, from chapter 8, with different step sizes, $h = 1.0$ and $h = 0.1$. Results are plotted for the first 200 time units, and again for the last 200 time units for an integration time of 100,000 time units. Numerical integration of the Lotka–Volterra predator–prey equations can be problematic because of its neutral stability, and small numerical errors can behave like small fluctuations, bumping the system into new trajectories. The numerical integration algorithm must be quite good to maintain a constant amplitude and period of oscillation. Numerical integration using $h = 1.0$ (9.9A) has a big problem at the end of the run, specifically the amplitude of the cycle has shrunk. In contrast, reducing h to 0.1 alleviates this problem. Problems like these are to be expected in numerical integration, and argue for seeking a convergent solution as one decreases the integration step size.

There is also a difference in period between the two integration timespans, demonstrating that the frequency depends on the amplitude for large amplitude oscillations. This difference is more easily seen in figure 9.9C, which plots the exact solution and the results for the final 200 time steps for both numerical integrations. Results for the short time step and the exact solution lie directly on top of one another, whereas the inner loop is the result for the long time step. Also notice that the cycle is not quite

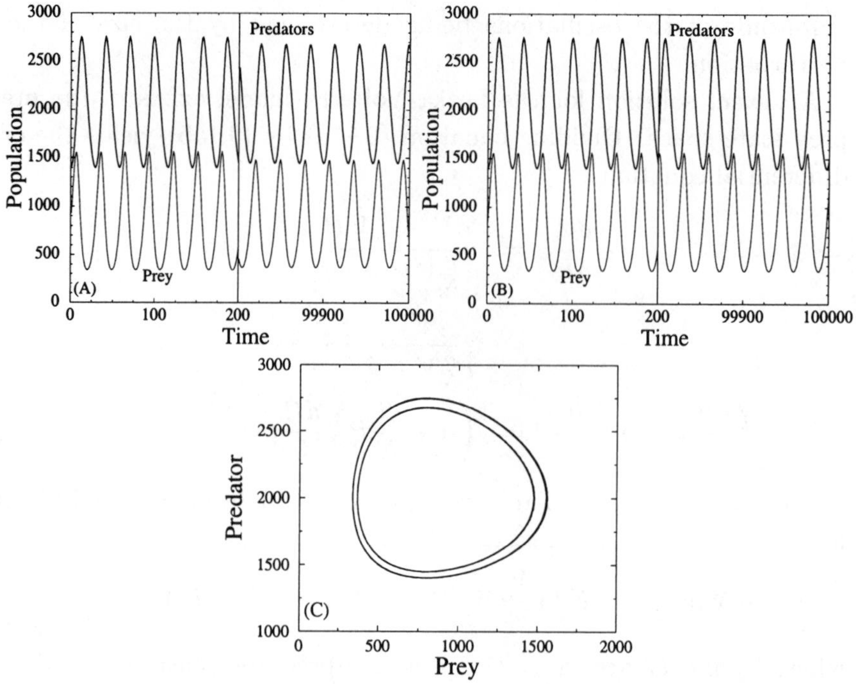

Figure 9.9. Numerical integration results for the Lotka–Volterra predator–prey model. (A) Integration step size $h = 1.0$. (B) Step size $h = 0.1$. Differences in the amplitudes and periods are evident. (C) Comparisons in the V–P plane shows the end results from $h = 0.1$ overlaps with the exact solution (outer cycle), but the cycle for $h = 1.0$ has shrunk.

circular; these results, taken together, indicate that the small perturbation assumption leading to equation (9.19a) is violated.

9.3.1.2 Spatial Model Analysis

A common way to extend models of interacting species into a spatial realm is by adding diffusion, yielding a reaction–diffusion model. For the Lotka–Volterra model it means making the prey and predator densities spatially explicit, $V(x,t)$ and $P(x,t)$, yielding the partial differential equations

$$\frac{\partial V}{\partial t} = \alpha V - \frac{\beta}{N} VP + D\frac{\partial^2 V}{\partial x^2} \tag{9.22a}$$

$$\frac{\partial P}{\partial t} = \epsilon \frac{\beta}{N} VP - \delta P + D\frac{\partial^2 P}{\partial x^2}. \tag{9.22b}$$

Here we assume that both species move across the same spatial scales in a fixed amount of time (equal diffusion constant D for both species). Reaction–diffusion models have a huge history in many fields, including predator–prey theory (e.g., Hadeler, an der Heiden, and Rothe 1974, McLaughlin and Roughgarden 1991, Holmes *et al.* 1994).

Equilibrium analysis of our reaction–diffusion model follows the same development as in the nonspatial model. First, when all derivatives are set to zero, the equilibria are unchanged with the addition of space. Stability analysis follows that of section 9.1.1; we consider the temporal evolution of spatial perturbations using trial solutions of the form

$$V(x,t) = V^* \left(1 + v_0 e^{-\lambda t} \cos(kx)\right) \tag{9.23a}$$

$$P(x,t) = P^* \left(1 + p_0 e^{-\lambda t} \cos(kx)\right). \tag{9.23b}$$

Replacing (9.23) into (9.22) yields a pair of equations that must be satisfied if the trial solution is a valid one

$$(\lambda - Dk^2)v_0 = \alpha p_0 \tag{9.24a}$$

$$-\delta v_0 = (\lambda - Dk^2)p_0. \tag{9.24b}$$

Dividing one equation by the other (setting the ratio of the left-hand sides equal to the ratio of the right-hand sides) yields the condition

$$(\lambda - Dk^2)^2 = -\delta\alpha, \tag{9.25}$$

and upon rearrangement gives the dispersion relation

$$\lambda = \sqrt{-\delta\alpha} + Dk^2. \tag{9.26}$$

Thus, the time-development of the perturbation has the functional form

$$\exp\left(-Dk^2 t\right) \cos\left(\omega t\right). \tag{9.27}$$

The first factor is identical to the expression we derived for the decay of perturbations in the pure diffusion model. The second factor makes use of the relationship for complex numbers, $\exp(i\theta) = \cos\theta + i\sin\theta$, where i is the imaginary number $\sqrt{-1}$, we discard the imaginary component, and $\omega = \sqrt{\delta\alpha}$ as in the nonspatial case.

The usual, deterministically motivated interpretation of these results is that if a system is initialized with a broad assortment of spatial perturbations, all spatial structure will eventually decay away leaving only the initial spatially homogeneous perturbation oscillating in a way that looks identical to the oscillations of the nonspatial model. However, long wavelength perturbations (those with $k \approx 0$) will take a long time (roughly $1/Dk^2$) to dissipate. In the following sections I demonstrate how these linear stability predictions, strictly applicable only to the deterministic model, break down for a simulation of prey and predators.[7]

Going beyond the stability analysis of deterministic equations, there are a variety of approaches. One approach is to make clever approximations that provide useful analytic results (Chesson 1981, Holmes *et al.* 1994). While accurate numerical integrations are important (e.g., Lewis 1994), another useful approach discretizes space[8] motivated by metapopulation structure (Hilborn 1975, Zeigler 1977, Crowley 1981, Comins, Hassell, and May 1991, Rohani and Miramontes 1995, Wilson *et al.* 1999).

9.3.2 Simulation Code Modifications

Few substantive changes from the code of the previous section are needed to simulate the predator–prey system described above. The primary change involves enhancements to the `Interact()` routine, reproduced here.

```
        void Interact(Site *arena, int *nprey, int *npred)
        {
            int  i, dir, flag, bcnt;
            Site *origin, *nnabe;

            origin = arena;              /* prey reproduction */
1.a)        for(i=0;i<NSITES;i++,origin++)
            {
                if(origin->prey==1)
                {
                    if(drand48()<alpha)
                    {
                        nnabe = origin;
                        flag = bcnt = 1;
2.a)                    while(flag==1 && bcnt<=10)
                        {
                            dir = 4*drand48();
                            nnabe = nnabe->nabes[dir];
                            if(nnabe->prey==1 || nnabe->vbabe==1)
                            {
                                bcnt++;
                            }
                            else
                            {
                                nnabe->vbabe = 1;
                                flag = 0;
```

7 Detailed analyses demonstrate useful applications of linear stability analyses of deterministic models for the interpretation of stochastic model results (Taylor 1992, Wilson 1998).

8 See Euler's method for discretizing time on page 186 – spatial discretization replaces spatial derivatives in a similar manner.

```
                    }
                }
            }
        }
    }
                /* predation and predator reproduction */
    origin = arena;
1.b)    for(i=0;i<NSITES;i++,origin++)
    {
        if(origin->prey * origin->pred==1)
        {
            origin->prey = 0;
            if(drand48()<epsilon)
            {
                nnabe = origin;
                flag = bcnt = 1;
2.b)                while(flag==1 && bcnt<=10)
                {
                    dir = 4*drand48();
                    nnabe = nnabe->nabes[dir];
                    if(nnabe->pred==1 || nnabe->pbabe==1)
                    {
                        bcnt++;
                    }
                    else
                    {
                        nnabe->pbabe = 1;
                        flag = 0;
                    }
                }
            }
        }
    }

1.c)    origin = arena;             /* predator death */
    for(i=0;i<NSITES;i++,origin++)
    {
        if(origin->pred==1)
        {
            if(drand48()<delta)
                origin->pred = 0;
        }
    }

    *nprey = *npred = 0;
    origin = arena;
1.d)    for(i=0;i<NSITES;i++,origin++)
    {
        if(origin->vbabe==1)
        {
```

```
                origin->vbabe = 0;
                origin->prey = 1;
            }
            if(origin->pbabe==1)
            {
                origin->pbabe = 0;
                origin->pred = 1;
            }
            if(origin->prey) (*nprey)++;
            if(origin->pred) (*npred)++;
        }
    }
```

9.3.3 Code Details

General comments: I added two new members to the `Site` structure, `vbabe` and `pbabe`, to hold the prey and predator offspring while running through the interactions between adults. Also, I defined `alpha`, `delta`, and `epsilon` globally, and modified `ArgumentControl()` to process `alpha`.

Line 1: The routine is broken into four basic parts, (a) prey reproduction, (b) predation and predator reproduction, (c) predator death, and (d) placement of offspring into the adult class. Notice that the specific order allows, for example, all prey entering the routine full opportunity to reproduce, independent of whether they are destined to be eaten by a predator. This careful specification keeps the different interactions from interfering with one another – assuring this independence has been called a concurrent implementation (McCauley, Wilson, and de Roos 1993). If a benchmark with an analytic model is the goal, less careful implementations introduce second-order interactions that produce contrasting results from those predicted by continuous-time models.[9]

Line 2: These two program sections make repeated attempts to place offspring into empty sites near the parent's site. Assuming that the overall reproduction rate is proportional to ρ, the species density per cell, and if only one attempt at placing offspring is made, then the resulting expression for population growth is logistic growth, $\rho(1-\rho)$, the product of offspring production and successful offspring placement. Above, an offspring

[9] In the context of a discrete-time analytic formulation, second-order interactions depend on Δt^2. In a simulation, second-order refers to two things happening within a cell during a single time step. Such interactions might be important biologically, but you should incorporate them intentionally, not accidentally!

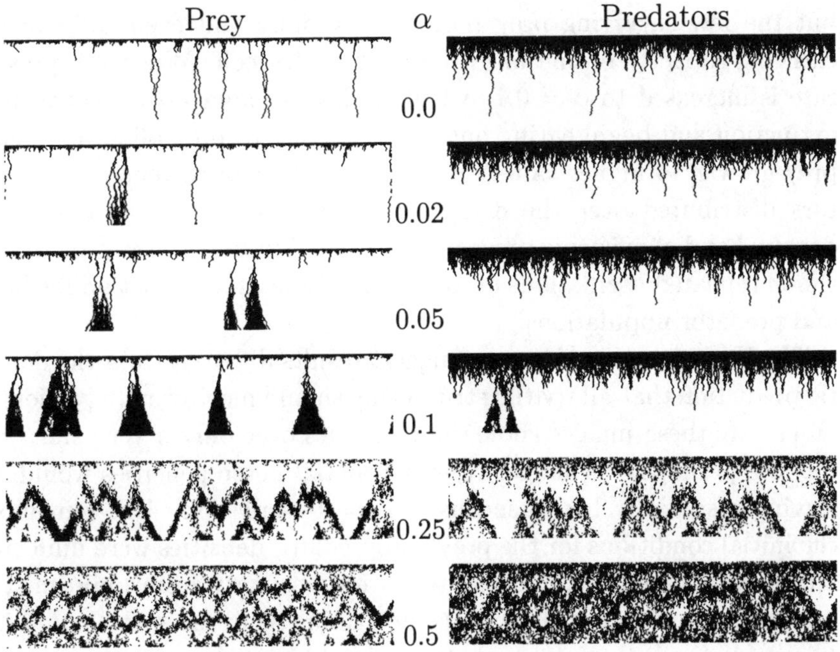

Figure 9.10. Prey and predator space–time images for several prey growth rates. Predators are unable to persist when the prey growth rate is too low (see exercise 9.11), but at a high-enough value both species persist. (All runs use $D = 1$, $\delta = 0.1$, and $\epsilon = 0.5$.)

tests up to ten sites before it perishes; if all attempts were independent of one another, then the resulting expression would be something like $\rho(1 - \rho)$.[10]

9.3.4 Simulation Results

The first set of results in figure 9.10 show the progression from a limit approximating the mutually annihilating particles of the previous section to the full Lotka–Volterra system of prey and predators. The key variable in this progression is the prey growth rate which changes from a value of $\alpha = 0.0$ to $\alpha = 0.5$ in the series. In the first four pairs of images, a pixel is shown as dark if any of the eight cells for a particular column in the lattice is occupied by the noted species, enabling better visualization of the interactions. In the last two pairs a pixel is shown as dark if a specific row's cell is occupied.

In the case of no prey growth, $\alpha = 0.0$, a few prey manage to hang on during the time it takes for the predators to go essentially extinct. When the prey growth rate takes on small values, the predators still face extinction, but the few surviving prey reproduce happily and eventually reach their carrying capacity of one individual per lattice cell. When the prey growth rate is increased to $\alpha = 0.1$, a few predators manage to survive the initial extinction and begin eating out the new pulse of prey offspring. Increasing prey growth to higher values, $\alpha = 0.25$, enables persistence of many predators distributed over the entire lattice and produces the observed high-density band of offspring predators. The final value for prey growth, $\alpha = 0.5$, shows repeated oscillations of localized growth and extinction for both prey and predator populations.

The last two pairs of images suggest results different from the deterministic prediction that all spatial structuring should melt away (equation (9.27)). Obviously these images show the dynamics over only a very short time interval – perhaps over longer runs the spatial heterogeneities would smooth, leaving a spatially homogeneous temporal oscillation. But remember that the initial conditions for the prey and predator densities were uniform sprinklings over the lattice, and any perturbations away from uniformity should have been small. According to the deterministic linear spatial stability analysis predictions, these perturbations should have decayed away.

Differences between the simulation results and deterministic predictions strike closer to the fundamental assumptions of deterministic modeling. Inherent in the process of writing down any partial differential equation, such as the diffusion equation, is the assumption that any infinitesmally small spatial region contains a large number of particles. In other words, stochastic variability is small only when the number of individuals in a region of arbitrary size is large. We derived this result in chapter 3 for the single-species logistic growth model (see equation (3.29)). This assumption of large populations in small regions does not hold in the prey–predator simulation, thus we run into stochasticity-induced differences.

We can test the importance of individuals and the impact of the assumption of large local populations by a simple modification of the simulation. The simulation used a thin, long lattice approximating a one-dimensional system – presumably the variation across the narrow width could be ignored. What we want to do now is increase the system's width, thereby increasing the number of individuals at each position along the length of the lattice, but assure no spatial heterogeneity across the width (Wilson 1998). As we increase the width the effect of local population size on the spatiotemporal dynamics will become clear.

At the heart of this test is a function that mixes the prey and predator populations over the width of the lattice. For example, the lines mixing the prey population are

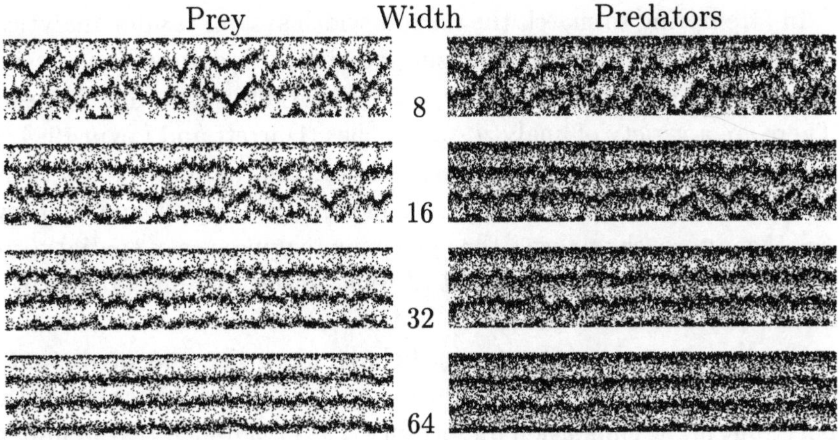

Figure 9.11. Mixing prey and predator populations across the width of a lattice allows the width to be treated like the size of local populations. When the local population size is large, the simulation results reproduce deterministic expectations of no spatial structure in the Lotka–Volterra predator–prey model.

```
newi1 = NROWS*drand48();
newi2 = NROWS*drand48();
temp = arena[newi1][j].prey;
arena[newi1][j].prey = arena[newi2][j].prey;
arena[newi2][j].prey = temp;
```

All we do is repeatedly exchange the prey states between two randomly chosen rows, `newi1` and `newi2`, for each column `i`. Figure 9.11 shows the effect of local population size by using a variety of system widths for $\alpha = 0.5$. When the system width is 8, we reproduce the comparable results of the unmixed system shown in figure 9.10. However, as the system width is increased the oscillations over the entire lattice length become synchronized in agreement with the expectations of equation (9.26).

The foregoing simulation model matched the inherent assumptions of the deterministic modeling framework by having large local populations and thereby recovered the predictions of spatially homogeneous oscillations. Thus, the source of the spatiotemporal patterning seen in the narrow width simulation model is due to stochastic processes highlighted in an individual-based simulation with small local population sizes. In spite of the differences between the two modeling frameworks seen for small local populations, linear spatial stability analysis provides useful guidance for understanding spatiotemporal patterns arising in more biologically realistic prey–predator simulation models (Wilson 1998).

In attempting to match the narrow width system results analytically, one moves into the realm of analyzing stochastic individual-based simulation models (e.g., de Roos, McCauley, and Wilson 1991, Keitt and Johnson 1995). There are a variety of analytic approaches (Durrett and Levin 1994, Boon *et al.* 1996). One approach uses stochastic calculus (Abundo 1991), which is an extension of stochastic birth–death models (see Nisbet and Gurney 1982). Another approach analyzes the approximate dynamics of spatial correlations (Matsuda *et al.* 1992, Harada *et al.* 1995, Rand, Keeling, and Wilson 1995, Satulovsky and Tomé 1997).

Finally, do ecological systems behave more like the predictions of linear stability analysis of deterministic equations, or are the stochastic effects borne out in the simulation important? That is a difficult, system-dependent question. As a rule-of-thumb, I postulate that if within one individual's lifetime spatial range there are a few hundred individuals with which it potentially interacts, then the population dynamics should follow deterministic predictions. If the number of individual is less than one hundred, then stochastic effects should be important.

9.4 Exercises

9.1 Equation (9.7) relates the wavelength of a perturbation, L, the diffusivity D of the organisms, and the relaxation timescale T. Test the validity of this equation. Initialize a one-dimensional lattice with alternating high and low density bands (the distance between successive high density bands is L). Write a measurement routine that measures the average density within a small set of cells, say of order $L/4$. Your starting distribution will give some initial difference between the highest measured density and the lowest measured density. Use your measurement routine to determine the time it takes for this difference to decrease to, say, one-fourth of its initial value. Do your measured times depend on L and D as expected?

9.2 Use offsets to write the diffusion routine. Then instead of passing a flag to move either prey or predators, pass the offset from the pointer to a structure variable for the prey or predator member. In this case, the structure pointer plus the offset locates the relevant variable.

9.3 Determine the equilibrium predator and prey densities of the Lotka–Volterra model. Derive the stability equations (9.17a) and (9.17b).

9.4 Derive equation (9.21).

9.5 Write a diffusion algorithm that specifies the particles' stepping order as their order in the `habitat` structure array and reproduce figure (9.5). Does everything look OK?

9.6 Decipher the following lines in `migrate.c`, alluded to on page 223, and explain how they work.

```
xxx = drand48()-bias[0];  /* choose step direction */
dir = 0; while(xxx>0.0) xxx -= bias[++dir];
nnabe = origin->nabes[dir];
```

9.7 Extend the two-species mutual annihilation simulation to a three-species mutual annihilation in which individuals of three different species must come together for an interaction to occur. Perform a run where individuals of all three species initially are randomly distributed over space.

9.8 Check the relation $\exp(i\theta) = \cos\theta + i\sin\theta$ by examining the Taylor series expansions for $\exp(i\theta)$, $\cos\theta$, and $\sin\theta$. Note that $i^2 = -1$.

9.9 Write a program producing the data for the exact results (rather, numerical results) of the Lotka–Volterra cycle, equation (9.21).

9.10 Perform the spatial linear stability analyses on equations (9.12) and (9.14). What are the predictions, and do they match the simulation results of figures 9.7 and 9.8?

9.11 Is the lack of predator persistence in the low prey growth rate images of figure 9.10 a result of the initial distribution of prey and predators? Instead of uniformly spreading predators across the entire lattice, initialize a small central region with just a few predators. Do predators persist for low (but nonzero) prey growth rates?

10
Optimal Resource Allocation Schedules

Programming components covered in this chapter:
• structures with arrays
• genetic algorithm for optimization problems

All organisms, at some point in their life histories, are faced with the physiological (or behavioral) "decision" of committing a resource such as time or energy to one function or another (Alexander 1996). For example, parasitoids that utilize their hosts for both food and oviposition must allocate hosts to one function or the other (Collier, Murdoch, and Nisbet 1994), and fish allocate energy to growth and reproduction (Roff 1983). In this chapter I consider the allocation decisions plants face in partitioning the resource produced through photosynthesis to either the development of additional photosynthetic material or to reproduction (Bazzaz *et al.* 1987). These types of allocation models are often called *dynamic energy budget models* (Kooijman 1993; Ross and Nisbet 1990, Engen and Sæther 1994) and take in as much detailed physiology as the problem demands and the modeler desires.

Given a model of energy allocation, the organism with the best strategy is the one that dies with the most resource devoted to reproduction.[1] Over evolutionary time, natural selection is assumed to optimize reproduction by partitioning available resource into various compartments, for example the roots, shoots, and reproductive structures of plants, as a function of a plant's age. A typical optimal strategy commits early resource production entirely into photosynthetic material, yielding greater resource production. Near the end of the plant's life, the allocation switches entirely to reproduction. Within this general conceptual framework, optimal life history strategies for

[1] This definition of "best" is contingent on many important assumptions; Here I assume a subsequent optimal packaging of reproductive resource into successful offspring.

plants (or autotrophic organisms in general) have been examined for a variety of scenarios, including simple herbivory (King and Roughgarden 1982), and static and variable environments (Cohen 1967, Cohen and Parnas 1976, Parnas and Cohen 1976, Vincent and Pulliam 1980, Iwasa and Roughgarden 1984, Iwasa 1989).

This chapter examines resource allocation models, and their optimal strategies, using analytic and simulation methods. The analytic methods begin with simple maximization concepts and culminate with optimal control theory (Rosen 1967), sometimes called Pontryagin's Maximization Principle (Bulmer 1994).[2] At the core of the simulation model is the so-called *genetic algorithm*, based rather loosely on genetic principles, that provides good solutions to complex optimization problems (Holland 1975, Goldberg 1989, Sumida *et al.* 1990, Wilson, Laidlaw, and Vasudevan 1994, Toquenaga and Wade 1996).

10.1 Resource Allocation in Plants

Consider a hypothetical plant of age t composed of two parts, photosynthetic material (leaf biomass), $L(t)$, and reproductive material, $R(t)$. At a given age the plant produces resource in proportion to its leaf biomass, which must then be allocated one of two ways – towards additional leaves which can then produce even more resource, or into a reproductive pathway assumed to be a direct correlate of the plant's fitness.[3] Among many assumptions, the simple allocation model examined here assumes plants of this species have a fixed and definite lifetime, T, and that, as a species, they face no environmental variability over evolutionary timescales. Resource allocation is controlled by an allocation schedule, $u_R(t)$, defined as the fraction of resource produced at age t allocated to reproduction, which itself is the only aspect of the plant's life history subject to selection forces. Questions of optimal allocation schedules in perennial plants can also be examined (Pugliese 1988, Iwasa and Cohen 1989)

System dynamics are defined by the growth trajectory of an individual plant's components $L(t)$ and $R(t)$. In the system described above, we have

[2] Such optimization approaches are fraught with peril when applied to stochastic, complex biological systems (see the clear discussion by Mangel and Clark 1988, page 4). Stochastic dynamic programming provides a more generalizable nonsimulation benchmark for harder problems than the simple allocation problem I examine.

[3] Thus allocation of photosynthate works much like economics: Save now for enhanced gratification later or spend now for a lesser amount of total gratification. Much of the biological theory behind optimal allocation schedules is borrowed directly from economic models (Intriligator 1971).

two differential equations

$$\frac{dL(t)}{dt} = r(1 - u_R(t))L(t), \qquad (10.1a)$$

$$\frac{dR(t)}{dt} = r u_R(t) L(t), \qquad (10.1b)$$

where the parameter r represents resource production by leaf biomass and the subsequent conversion into new leaf biomass. Finally, let the plant's reproductive output integrated over its fixed lifetime, $R(T)$, be a measure of its fitness

$$R(T) = r \int_0^T u_R(t) L(t) dt. \qquad (10.2)$$

First we will examine a simulation of two allocation schedules, each defined by a single trait subject to optimization. These schedules provide clean connections between analytic and simulation models providing initial insight into allocation problems. Later we examine optimization of the most general allocation schedule. Analytic methods are robust enough and the model is simple enough to allow an exact analytic solution to the optimal allocation schedule. Results from a simulation implementing the genetic optimization algorithm are then compared with the analytically derived schedule.

10.2 Two Hypothetical Allocation Schedules

A plant might have one of two allocation strategies: a constant allocation throughout its lifetime, or a specific time in its life that it switches from pure growth to pure reproduction.[4] These schedules are depicted in figure 10.1. Both schedules allow integration of equation (10.1).

The first schedule, defined by a constant allocation $0 < c < 1$ throughout a plant's lifetime, gives the vegetative material at any time t

$$\frac{dL}{dt} = r(1 - c)L \implies L(t) = L(0)e^{r(1-c)t}. \qquad (10.3)$$

Replacing $L(t)$ into equation (10.1b)

$$\frac{dR(t)}{dt} = rcL(0)e^{r(1-c)t} \implies R(t) = \frac{cL(0)}{1-c}\left(e^{r(1-c)t} - 1\right), \qquad (10.4)$$

assuming an initial condition of no reproductive material, $R(0) = 0$.

The second schedule, defined by no reproductive allocation before a switch

[4] One is an obvious choice and the other is the solution to the general case.

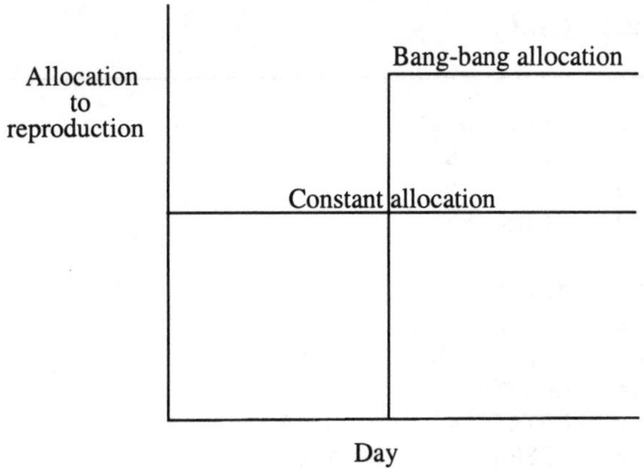

Figure 10.1. Two hypothetical resource allocation schedules for plants, one a constant allocation to reproduction throughout the entire season, the other a sudden switch from growth to reproduction. The latter schedule is commonly called a *bang–bang* schedule.

time t_s and full allocation to reproduction after t_s, also allows an exact determination of $L(t)$ and $R(t)$. Here we have at the switch time

$$L(t_s) = L(0)e^{rt_s} \tag{10.5}$$

($L(t_s)$ for the first schedule with $c = 0$), and at the end of the season

$$R(T) = rL(0)e^{rt_s}(T - t_s). \tag{10.6}$$

This schedule is called *bang–bang* because the allocation goes from one extreme to the other with no intermediate allocations. Following a look at a simulation modeling these two schedules, we derive this latter schedule with a specific optimal switch time.

10.2.1 Simulation of Hypothetical Schedules

Let's explore the results of an individual-based simulation that involves a spatially distributed plant population run over many generations. A grid of cells comprises the spatial structure, and each cell contains one plant. Each plant has a distinct allocation schedule defined by a plant-specific value for c or t_s that allocates photosynthetically produced material between vegetative and reproductive parts. At the end of a season reproduction occurs between plants, producing seeds with potentially mutated allocation parameters. These seeds are distributed over the habitat yielding the next generation of plants.

10.2.2 Code

```
    #include  <stdlib.h>
    #include  <stdio.h>
    #include  <math.h>
    #include  <time.h>

    #define   MAXGENS        100
    #define   NPLANTS        1000
    #define   SEASONLEN      100.0
    #define   GROWTH         0.02
    #define   MUTATERATE     0.05
    #define   LZERO          1.0
    #define   SOLNTYPE  1 /* 1-constant, 2-switch */

    typedef struct plant_def
    {
         double    trait;
         double    size;
         double    fitness;
    } Plant;
1.d) Plant plants1[NPLANTS], plants2[NPLANTS];

    void InitSeed(void);
    void InitPlants(Plant *plnt);
    void ZeroPlants(Plant *plnt);
    void GrowPlants(Plant *plnt);
    void ReproducePlants(Plant *parents, Plant *babies);
    void PrintPlants(Plant *plnt);

    int main(void)
    {
         int       ttt;
1.e)     Plant     *babes, *adults, *tmpy;

         adults = plants2; babes = plants1;
         InitSeed();
1.f)     InitPlants(babes);

         for(ttt=0;ttt<MAXGENS;ttt++)
         {
1.c)          tmpy = adults; adults = babes; babes = tmpy;
              ZeroPlants(adults);
1.a)          GrowPlants(adults);
1.b)          ReproducePlants(adults,babes);
         }
         PrintPlants(adults);
         return(0);
    }
```

```
void InitPlants(Plant *plnt)
{
     int i;

     for(i=0;i<NPLANTS;i++,plnt++)
     {
#if(SOLNTYPE==1)
          plnt->trait = drand48();
#endif
#if(SOLNTYPE==2)
          plnt->trait = drand48()*SEASONLEN;
#endif
     }
}

void ZeroPlants(Plant *plnt)
{
     int i;
     for(i=0;i<NPLANTS;i++,plnt++)
     {
          plnt->size = LZERO;
          plnt->fitness = 0.0;
     }
}

void GrowPlants(Plant *plnt)
{
     int i;
     double efac;

     for(i=0;i<NPLANTS;i++,plnt++)
     {
#if(SOLNTYPE==1)
          efac = exp(GROWTH*(1.0-plnt->trait)*SEASONLEN);
          plnt->size = LZERO*efac;
          plnt->fitness = (GROWTH*plnt->trait*LZERO)*(efac-1.0)
                              /(GROWTH*(1.0-plnt->trait));
#endif
#if(SOLNTYPE==2)
          plnt->size = LZERO*exp(GROWTH*plnt->trait);
          plnt->fitness = GROWTH*plnt->size
                    *(SEASONLEN-plnt->trait);
#endif
     }
}

void ReproducePlants(Plant *parents, Plant *babies)
{
     int  i, j;
```

```
          double     totrep, rnum;
          double     brakes[NPLANTS];
          Plant      *plnt;

          plnt = parents; /* get total fitness */
          totrep = 0.0;
2.a)      for(i=0;i<NPLANTS;i++,plnt++)
          {
               totrep += plnt->fitness;
               brakes[i] = totrep;
          }

          for(i=0;i<NPLANTS;i++,babies++) /* Clonal reproduction */
          {          /* choose adults proportional to fitness */
2.b)           rnum =  drand48() * totrep;

2.c)           j = (int)((rnum*NPLANTS)/totrep);
               if(rnum<brakes[0]) j = 0;
               else if(brakes[j]<rnum)
                    while(rnum > brakes[j]) j++;
               else
               {
                    while(rnum < brakes[j]) j--;
                    j++;
               }

2.d)           babies->trait = parents[j].trait;

2.e)           if(drand48()<MUTATERATE)
     #if(SOLNTYPE==1)
                    babies->trait = drand48();
     #endif
     #if(SOLNTYPE==2)
                    babies->trait = drand48()*SEASONLEN;
     #endif
          }
     }

     /* dist[] holds freq distribution of trait vals
        flag[] marks counted individuals
          (0-uncounted, -1 marks newtype, 1-counted) */
     void PrintPlants(Plant *plnt)
     {
          int i, j, flag[NPLANTS], dist[NPLANTS], biggie=0;

3.a)      for(i=0;i<NPLANTS;i++)
               flag[i] = dist[i] = 0;

3.b)      for(i=0;i<NPLANTS;i++)
          {
```

```
                if(flag[i]==0) /* uncounted */
3.c)            {
                        dist[i] = 1;
                        flag[i] = -1;
3.d)                    for(j=i+1;j<NPLANTS;j++) /* seek out clones */
                        {
                                if(plnt[i].trait==plnt[j].trait)
                                {
                                        dist[i]++;
                                        flag[j] = 1;
                                }
                        }
3.e)                    if(dist[i]>biggie) biggie = dist[i];
                }
        }

3.f)    for(i=0;i<NPLANTS;i++,plnt++) /* print unique clones */
        {
                if(flag[i]==-1)
                        printf("%f  %f  %f %f\n",
                                plnt->trait,plnt->size,plnt->fitness,
                                5.0*dist[i]/(double)biggie);
        }
}
```

10.2.3 Code Details

Line 1: This simulation has two essential parts: growth and reproduction. Each generation consists of determining the size and fitness of each plant within the function `GrowPlants()` at line (1.a) according to equations (10.3) and (10.4), or (10.5) and (10.6), dependent on the allocation schedule used (toggled using `SOLNTYPE`). Once all the plant fitnesses have been calculated, the simulation moves to `ReproducePlants()` at line (1.b). A new use of pointers that comes in very handy in this array updating situation is shown at line (1.c). Two plant arrays were declared at line (1.d), one used as an offspring array, the other as an adult array, but trading places every generation. Three `Plant` pointers are defined locally at line (1.e), two of which are initialized to point to the declared arrays. The offspring plants are initialized at line (1.f). Control then passes into the loop, representing a new generation. Offspring then become adults through a switch of pointers performed at (1.c). The old adult array now becomes the offspring array, ready to be reset during reproduction. Switching pointers as done at (1.c) is one of the beautiful features of C that drew me to the language when I learned it. Using pointers to swap arrays avoids having to copy values from one array to another just to use those values in a different context, copying that takes many, many CPU cycles.

Line 2: `ReproducePlants()` consists of two parts. First, the total population fitness is measured. A running total is kept as each plant is added, and I call the total after a plant is added the plant's breakpoint. These values are held in the array `brakes[]` at line (2.a). When producing offspring plants, I choose a fitness point between 0 and the total population fitness at line (2.b). At line (2.c) I search for the plant that corresponds to this point by first starting a proportional distance into the `brakes[]` array, then searching up or down the list until I find the chosen plant. At line (2.d) the baby inherits the parent's trait, unless the baby becomes a mutant at line (2.e).

Line 3: Data output is performed in `PrintPlants()`. Considering that there are many plants having many different trait values, it would be nice to see some of this distributional information. In this procedure I hold the frequency distribution in the array `dist[]`, and processing information in the array `flag[]`. These two arrays are zeroed at line (3.a). I then run through all the plants at (3.b). If the `i`th plant's flag is unset at (3.c), then its trait value has not been encountered previously, so I set its distribution array element to `1` and its flag to `-1`. The flag value will denote an array element with information to plot later. At line (3.d) I search all the remaining plants for clones of the `i`th plant, or those having identical trait values. Each time I find one I increment the `i`th plant's distribution counter and set the `j`th plant's flag to one. At (3.e) I find the highest clone population size and use this value to scale the distribution uniformly. Once all the plants have been counted, I run through the array and print out all the information.

10.2.4 Simulation Results

Figure 10.2 summarizes the simulation results for the two predefined schedules discussed above. In the case of a constant daily allocation throughout the season (panel A), the fitness appears maximized when all of the produced resource is allocated to reproduction. Both the size at the end of the season and the final fitness follow smooth functional forms, representing the deterministic functions underlying these quantities. The observed distribution of traits, however, is very broad and stochastic, reflecting the random processes of reproduction and mutation in light of a broad fitness peak.

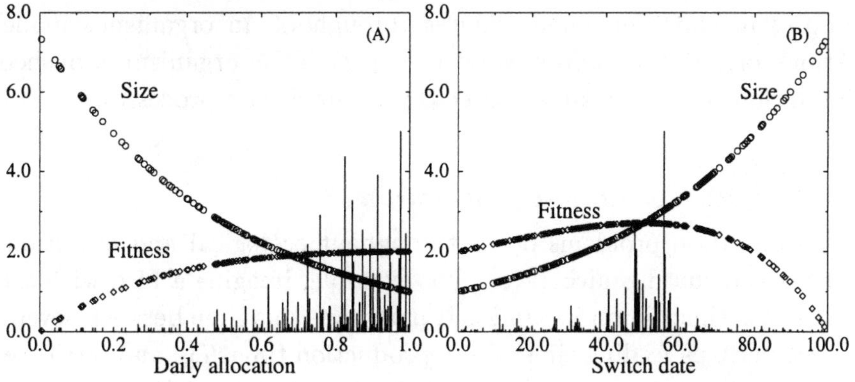

Figure 10.2. Results from simulations of the two hypothetical plant allocation strategies, (A) constant allocation and (B) bang–bang allocation. Vertical lines represent the relative frequencies of plants having a particular trait value, either the constant daily allocation or the switch date. For each trait value having at least one individual represented, the size and fitness at season's end is plotted. These latter curves plot the deterministic curves (equations (10.3)–(10.6)).

Daily allocation of all resource to reproduction in panel (A) corresponds to switching allocation to reproduction on the first day in panel (B). We can observe this equality by the way the fitness and size curves match at the upper and lower trait limits in the first and second cases, respectively, where the figures are closest. The fitness in the second case continues to increase with a maximum somewhere around day 50.

10.3 Optimal Allocation Schedules

The previous simulation compared two reasonable, hypothetical schedules for the way plants, or other similar autotrophic organisms, might allocate their produced resource between growth and reproduction. In this section I discuss deriving optimal allocation schedules from both a mathematical perspective and a simulation approach implementing the so-called genetic algorithm. I first introduce an analytic method for calculating optimal trait values in static (time-independent) optimization problems. I then generalize the method developed for static problems to dynamic (time-dependent) problems, yielding a method closely related to dynamic programming approaches (Schaffer 1983, Stephens and Krebs 1986, Mangel and Clark 1988), and derive the optimal allocation schedule for autotrophic organisms. The deterministic approach I discuss produces the optimal path from start to finish, whereas stochastic dynamic programming results yields optimal, state-dependent choices throughout an organism's modeled lifetime, or, in other words, what to expect if the organism is pushed off the deterministically optimal trajectory by stochastic processes.

10.3.1 Static Optimization

Optimization problems occur throughout ecological and evolutionary systems (Maynard Smith 1978). For example, imagine a bird with a territory that has time, T, on its wings. It partitions this time between several activities, perhaps feeding time, T_F, reproduction time, T_R, and territory defense time, T_D. Selection should favor behaviors that maximize reproductive output by optimally dividing the bird's time between these three important activities. Presumably, it could devote all of its time to reproduction, but it is prevented from doing so for at least two reasons. First, if it doesn't eat, it dies, and death puts a crimp on mating opportunities. Likewise, without defense it will lose its territory and assume, for the sake of argument, that no territory means no mating opportunities. Thus, our hypothetical bird's net fitness depends on these various times in a complicated manner. Further, it cannot allocate more time to these activities than it has: $T = T_R + T_F + T_D$. This latter expression is called a constraint.

There are at least two static optimization methods allowing for the automatic incorporation of constraints within the solution for the optimum – direct substitution and Lagrange multipliers (e.g., Lloyd and Venable 1992). As a pedagogical example, consider maximizing the fitness function

$$W(x, y) = C - x^2 - y^2. \tag{10.7}$$

The optimal value for both x and y is zero, for which $W(0,0) = C$. However, suppose that a constraint exists between x and y

$$y = a + bx, \tag{10.8}$$

where a and b are fixed constants unaffected by evolutionary processes. In a biological process this constraint might represent energetic, time, or genetic tradeoffs. Maximizing $W(x, y)$ is now a little bit harder. The situation is depicted in figure 10.3.

10.3.1.1 Directly incorporating constraints.

The most direct and obvious way of maximizing $W(x, y)$ given the constraint, equation (10.8), is by deriving the dependence of one variable on the other via the constraint, plugging that into the expression to be maximized,

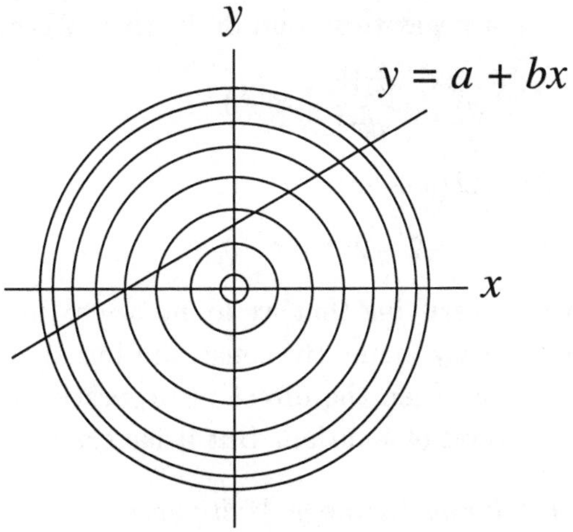

Figure 10.3. Representation of the fitness surface for equation (10.7) with the maximum at the origin. The concentric circles represent lines of equal suboptimal fitness. A constraint, $y = a + bx$, restricts the system to exist on the line slicing through the (x,y) trait space.

then maximizing. I intentionally chose a particularly simple constraint of the form

$$y = f(x), \tag{10.9}$$

that allows direct substitution into the fitness function, $W(x, f(x))$. In this case, direct substitution yields

$$W(x, f(x)) = C - x^2 - (a + bx)^2, \tag{10.10}$$

and we obtain the optimal value x^* from

$$\frac{dW}{dx} = 0. \tag{10.11}$$

We can solve this expression for the optimal value x^*

$$-2x^* - 2b(a + bx^*) = 0 \tag{10.12a}$$

$$x^* = \frac{-ab}{(1 + b^2)}. \tag{10.12b}$$

The constraint forces the maximum to occur on the line $y = f(x)$, and the optimization process finds the maximum value of W on that line. Technically, we still need to see if this optimal point is really a maximum and not a minimum. Remember, a maximum looks like a hump, not like a valley, and humps have a negative second derivative. We want to check whether

$$\frac{d^2W}{dx^2} < 0, \tag{10.13a}$$

$$\frac{d}{dx}(-2x - 2b(a + bx)) < 0, \tag{10.13b}$$

$$-2 - 2b^2 < 0, \tag{10.13c}$$

which is satisfied, verifying that we found a maximum.

If it was always easy to find the constraint function $f(x)$, and it was always a "pretty" function, then the direct incorporation of constraints would be the preferred method of solution. But it isn't always so.

10.3.1.2 Using Lagrange Multipliers.

Another approach to solving these static optimization problems is to first make the problem look more difficult through the use of things called *Lagrange multipliers.* Lagrange multipliers allow you to incorporate the constraint(s) within a revised fitness function while treating the Lagrange multiplier as a new optimized parameter. Using them to solve a constrained optimization problem gives us a revised quantity to optimize

$$H(x, y, \lambda) = W(x, y) + \lambda g(x, y). \tag{10.14}$$

λ is the Lagrange multiplier associated with the constraint $g(x, y) = 0$, and H is called the Hamiltonian function. Adding the product of this multiplier and the constraint (set up to equal 0) to the original fitness function, yields the Hamiltonian (which is not a fitness function). We now have three quantities whose optimal values, x^*, y^*, and λ^*, must be found. This solution is accomplished via the standard maximization route of solving the three simultaneous equations

$$\frac{\partial H}{\partial x} = 0, \tag{10.15a}$$

$$\frac{\partial H}{\partial y} = 0, \tag{10.15b}$$

$$\frac{\partial H}{\partial \lambda} = 0. \tag{10.15c}$$

The best way to think about λ is that it is a quantity whose value is chosen to satisfy equation (10.15b) for all values of x^* and y^*, irrespective of the fact that those two values are constrained. Equation (10.15c) simply gives back the constraint, $g(x, y) = 0$, providing the third equation (independent of λ) for the system of three unknowns. Hence, this procedure turns a problem of optimizing two unknowns linked by one constraint into a problem of optimizing three unconstrained unknowns. Often this latter problem is easier than the former. Also, this procedure is entirely generalizable to optimization problems with multiple constraints.

Rather than actually justifying this approach through rather convoluted mathematics, I demonstrate it for the optimization example solved above by the direct method. For that problem, we have the Hamiltonian

$$H(x, y, \lambda) = C - x^2 - y^2 + \lambda\{y - (a + bx)\}, \tag{10.16}$$

and optimizing it leads to the three equations

$$\frac{\partial H}{\partial x} = 0 \rightarrow -2x^* - b\lambda^* = 0, \tag{10.17a}$$

$$\frac{\partial H}{\partial y} = 0 \rightarrow -2y^* + \lambda^* = 0, \tag{10.17b}$$

$$\frac{\partial H}{\partial \lambda} = 0 \rightarrow y^* - (a + bx^*) = 0. \tag{10.17c}$$

Solving (10.17c) gives us the original constraint, $y^* = a + bx^*$. Replacing this expression into (10.17b) gives $\lambda^* = 2(a + bx^*)$. Finally, replacing this expression into (10.17a)

$$x^* = \frac{-b\lambda^*}{2} \tag{10.18a}$$

$$= \frac{-b2(a + bx^*)}{2} \tag{10.18b}$$

$$\Longrightarrow x^* = \frac{-ab}{1 + b^2}. \tag{10.18c}$$

As expected, this result is identical to the direct method's result. Unfortunately, the test of whether this value corresponds to a maximum rather than a minimum is difficult (Intriligator 1971).

10.3.2 Dynamic Optimization

Dynamic optimization is a problem common to many disciplines, engineering, physics, mathematics, economics, and biology, yielding solution methods given a variety of names including control theory, calculus of variations, Pontryagin's Maximization Principle, and dynamic programming (Intriligator 1971, Mangel and Clark 1988, Perrin and Sibly 1993). An excellent discussion is given by Alexander (1996).

The problem was outlined in section 10.2, but let's review it here in light of the previous discussion on static optimization. Our plant, composed of photosynthetic material, $L(t)$, and reproductive material, $R(t)$, produces resource with rate $rL(t)$. A schedule, $u_R(t)$, controls the allocation of resource production into reproduction, with the plant's fitness being the allocation to reproduction integrated over its fixed lifetime T. Hence, the plant's fitness is

$$R(T) = \int_0^T u_R(t)L(t)dt. \tag{10.19}$$

The plant's schedule is constrained in that the photosynthetic material obeys the differential equation

$$\frac{dL(t)}{dt} = r(1 - u_R(t))L(t). \tag{10.20}$$

Equation (10.20) denotes that the increase in photosynthetic material is constrained by the amount left over after a portion of the resource is allocated to reproduction.

For greater generality in the analytical derivation of the optimal allocation schedule, we can write these equations as

$$R(T) = \int_0^T I(u_R, L)dt \tag{10.21}$$

subject to the constraint

$$\frac{dL}{dt} = f(u_R, L). \tag{10.22}$$

The constraint holds at every instant in time.

One way to think about this constraint is that it is actually composed of many constraints, one for each point in time. Maximizing $R(T)$ is akin (but not equivalent) to maximizing reproduction at each of these points subject to a time-specific constraint. Each of these time-specific optimizations behaves just like a static optimization problem with a constraint

$$J_i\Delta t = I_i(u_R, L)\Delta t + \lambda_i\Delta t\left(f(u_R, L) - \frac{dL}{dt}\right), \tag{10.23}$$

where we imagine the season of length T broken into short time intervals of duration Δ_t. J_i is then similar to the Hamiltonian function for the ith time interval. If we only had to deal with a single point in time, we know how to do this optimization from the above section. However, during each time interval one of these static optimizations must be done, and the results of one interval fed into the initial conditions of the next interval. Thus the goal of optimizing reproduction over the whole season, T, turns into optimizing a functional J, where

$$J = \sum_i J_i \Delta t \tag{10.24a}$$

$$J = \sum_i \left\{ I_i(u_R, L) + \lambda_i \left(f(u_R, L) - \frac{dL}{dt} \right) \right\} \Delta t \tag{10.24b}$$

$$= \int_0^T \left\{ I(u_R, L) + \lambda \left(f(u_R, L) - \frac{dL}{dt} \right) \right\} dt. \tag{10.24c}$$

In the last line I let $\Delta t \to 0$ yielding the integral over time, and λ is implicitly a function of time, $\lambda = \lambda(t)$. This limiting procedure recovers our continuous-time formulation, equations (10.21) and (10.22), except with the constraint incorporated into the integral.

In the quest of maximizing J, we can apply the identity for the integration of a product of two functions, called integration by parts. This identity makes use of the equality

$$\int_0^T \frac{d(xy)}{dt} dt = \int_0^T \left(y\frac{dx}{dt} + x\frac{dy}{dt} \right) dt, \tag{10.25a}$$

which upon integrating the left-hand side becomes

$$x(T)y(T) - x(0)y(0) = \int_0^T y\frac{dx}{dt} dt + \int_0^T x\frac{dy}{dt} dt \tag{10.25b}$$

$$\int_0^T x\frac{dy}{dt} dt = x(T)y(T) - x(0)y(0) - \int_0^T y\frac{dx}{dt} dt. \tag{10.25c}$$

Applying integration by parts to the third term in equation (10.24c), identifying $\lambda(t)$ as x and L as y, gives

$$J = \int_0^T \left\{ I(u_R, L) + \lambda f(u_R, L) + \frac{d\lambda}{dt} L \right\} dt - \{\lambda(T)L(T) - \lambda(0)L(0)\}. \tag{10.26}$$

Defining the Hamiltonian for this problem

$$H(u_R, L, \lambda) = I(u_R, L) + \lambda f(u_R, L), \tag{10.27}$$

we can rewrite the functional as

$$J = \int_0^T \left\{ H(u_R, L, \lambda) + \frac{d\lambda}{dt} L \right\} dt - \{\lambda(T)L(T) - \lambda(0)L(0)\}. \tag{10.28}$$

If we possessed the optimal schedule, u_R^*, and resultant leaf biomass trajectory, L^*, such that $R(T) = R^*(T)$ and $J = J^*$ were maximized, *then* small variations about u_R^* and L^* (call them Δu_R and ΔL) would not alter $R(T)$. This assertion is just the definition of a maximum, with the important exception that a variation might increase $R(T)$ but the variation is disallowed by a constraint.[5] One constraint, for example, is that the plant cannot allocate more resource to reproduction than is actually produced, $u_r \leq 1$.

As done in a stability analysis, we substitute $u_R = u_R^* + \Delta u_R$ and $L = L^* + \Delta L$ into equation (10.28) and Taylor series expand

$$\begin{aligned} J &= \int_0^T \left\{ H(u_R^* + \Delta u_R, L^* + \Delta L, \lambda) + \frac{d\lambda}{dt}(L^* + \Delta L) \right\} dt \\ &\quad - \{\lambda(T)[L^*(T) + \Delta L(T)] - \lambda(0)L(0)\} \qquad (10.29a) \\ J^* + \Delta J &= \int_0^T \left\{ \left[H(u_R^*, L^*, \lambda) + \frac{\partial H}{\partial u_R}\Delta u_R + \frac{\partial H}{\partial L}\Delta L \right] \right. \qquad (10.29b) \\ &\quad \left. + \frac{d\lambda}{dt}(L^* + \Delta L) \right\} dt - \{\lambda(T)[L^*(T) + \Delta L(T)] - \lambda(0)L(0)\} \\ \Delta J &= \int_0^T \left\{ \frac{\partial H}{\partial u_R}\Delta u_R + \left(\frac{\partial H}{\partial L} + \frac{d\lambda}{dt} \right) \Delta L \right\} dt \\ &\quad - \lambda(T)\Delta L(T). \qquad (10.29c) \end{aligned}$$

Although messy, the meaning of this last expression is straightforward and very helpful. First of all, because the optimal schedule is optimal, any small variation about the schedule has to yield $\Delta J = 0$. Next, we could consider variations Δu_R and ΔL that are completely arbitrary and independent from one another, or, in other words, for any particular choice of Δu_R we could consider many different choices for ΔL. Since $\Delta J = 0$ for all these different combinations, the implication is that the coefficients on Δu_R, ΔL, and $\Delta L(T)$ must all be separately zero for all times. These coefficients yield three expressions that determine the optimal allocation schedule

$$\begin{aligned} \frac{\partial H}{\partial u_R} &= 0, \qquad (10.30a) \\ \frac{d\lambda}{dt} + \frac{\partial H}{\partial L} &= 0, \qquad (10.30b) \\ \lambda(T) &= 0. \qquad (10.30c) \end{aligned}$$

These expressions also look rather forbidding, but in fact are quite useful.

Returning to our specific plant resource allocation problem we can combine equations (10.19), (10.20), and (10.27), for which the Hamiltonian is

[5] This process is sometimes called variational calculus.

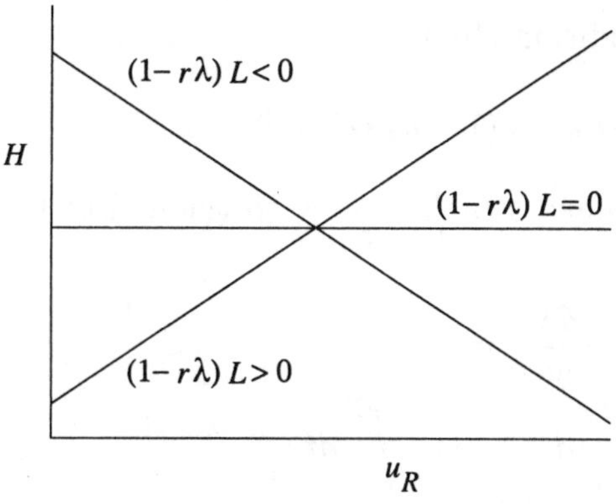

Figure 10.4. Dependence of H on the allocation parameter u_r (equation (10.31)) for the three possible cases. These dependencies imply that u_R will only take the extreme values, $u_R = 0$ or $u_R = 1$.

$H = u_R L + r\lambda(1 - u_R)L$, whereby condition (10.30a) gives

$$\frac{\partial H}{\partial u_R} = (1 - r\lambda)L = 0. \tag{10.31}$$

At first glance, this expression seems to say only that $\lambda(t) = 1/r$, which would seemingly mean that the function $\lambda(t)$ is a constant value for all time. This conclusion is not correct; equation (10.30b) determines $\lambda(t)$. Condition (10.31)'s interpretation requires more thought. The expression $\partial H/\partial u_R$ represents the maximization of H with respect to the reproductive allocation u_R, but the right-hand side is independent of u_R. This situation gives the dependence of H on u_R shown in figure 10.4. If $(1 - r\lambda)L > 0$, $\partial H/\partial u_R$ is positive, it means that H increases forever with increasing u_R, and natural selection makes u_R as big as possible. On the other hand, if $(1 - r\lambda)L < 0$, $\partial H/\partial u_R$ is negative, meaning H decreases forever with increasing u_R and selection makes u_R as small as possible. Since $0 \le u_r \le 1$ are the limits on u_R, it takes on either the value 0 or 1 and will never take on an intermediate value. Only at one very specific crossover point does the condition hold exactly. The result is the bang–bang schedule (as hypothesized on page 251).

Note that since equation (10.30c) gives $\lambda(T) = 0$, then $(1 - r\lambda(T))L(T) = L(T) > 0$ means $\frac{\partial H}{\partial u_R} > 0$ at the end of the season. Thus, the plant allocates everything to reproduction at the end of the season, $u_R(T) = 1$. When did the plant start allocating all of its produced resource to reproduction?

Integrating condition (10.30b)

$$\frac{d\lambda}{dt} = -u_R + r(1 - u_R)\lambda, \tag{10.32}$$

from this presumed switch time t_s to the end of the season (during which $u_R = 1$)

$$\frac{d\lambda}{dt} = -1 \tag{10.33a}$$

$$\int_{\lambda(t_s)}^{\lambda(T)} d\lambda = -\int_{t_s}^{T} dt \tag{10.33b}$$

$$\lambda(T) - \lambda(t_s) = -(T - t_s). \tag{10.33c}$$

At t_s, the equality holds in equation (10.31) (the flat line in figure 10.4) giving $\lambda(t_s) = \frac{1}{r}$. Replacing $\lambda(t_s)$ into (10.33c) gives

$$-\frac{1}{r} = -(T - t_s) \Rightarrow t_s = T - \frac{1}{r}. \tag{10.34}$$

This is it! Substituting the values we used in section 10.2.1 for the bang-bang allocation schedule simulation, $T = 100$ and $r = 0.02$, the optimal switch time from pure growth to pure reproduction is $t = 50$. This time corresponds precisely to the optimum fitness observed in figure 10.2B.

10.4 Simulation of General Allocation Problem

The optimal allocation schedule was derived analytically above. Here I revise the previous simulation, treating the allocation between growth and reproduction for each day as a separate trait open to mutation and selection. My simulation represents a genetic optimization algorithm, but the variability between details of the algorithms is very high (Holland 1975, Goldberg 1989, Wilson, Laidlaw, and Vasudevan 1994). Another optimization method for complex problems is simulated annealing (Kirkpatrick, Gelatt, and Vecchi 1983, Vasudevan, Wilson, and Laidlaw 1991), a method based on the theory of phase transitions (e.g., ice melting or water freezing). Both methods are discussed by Bounds (1987).

10.4.1 Overview of Genetic Algorithms

Evolution is a process whereby the alleles of a biological species (with thousands of genes) arise, then wax and wane in response to the environmental features affecting fitness. In an optimization problem a specific performance function defines a constant environment; optimization's goal is choosing the

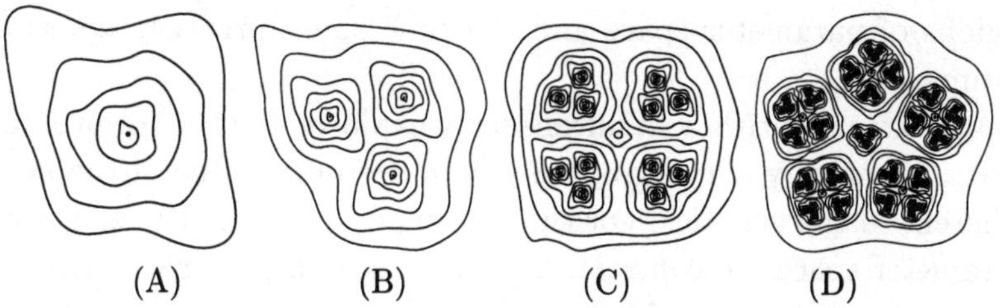

Figure 10.5. Increasingly rugged optimization surfaces. When the optimization surface is simple, trivial search algorithms will work well, but a complicated surface with many local maxima requires complicated search algorithms.

set of parameters yielding the highest performance. Exploiting the similarities with optimization problems, an abstraction of the evolutionary process has produced the so-called genetic algorithm, a demonstrably useful technique for complex problems.

Simple optimization problems don't require fancy techniques. Consider the optimization surface shown in figure 10.5A. An efficient search for the optimum in this case is a gradient search: Starting from a randomly chosen point in parameter space, measure the local gradient and move a short distance up the gradient to a higher performances. Repeat this procedure until the maximum is reached. A gradient search can be extended to a more complicated optimization surface (figure 10.5B): Randomly choose many starting positions and accept the highest peak from among those found as the global optimum.

The greatest challenge in optimization problems occurs when the number of parameters and the number of their possible values are large and the optimization surface, or fitness landscape, is as jagged as a mountain range (figure 10.5D). Gradient searches become impractical simply because the number of local maxima is too large – in the worst case every point must be examined. If there are 100 parameters with just ten states for each parameter, then the number of points in parameter space (unique parameter combinations) is 10^{100}. Searching this space at a rate of 10^{24} points/sec would require 10^{60} lifetimes of the universe[6] to search the entire phase space! Any reasonably efficient search method must ignore most of this parameter space and concentrate on the regions containing high performance values. Thus a tradeoff must be made between the time spent exploring the vast

[6] For scale, the lifetime of the universe is roughly 10^{17} seconds, whereas the time one spends in graduate school is roughly 10^{8} seconds.

regions of parameter space and the time spent precisely determining local optima.

A genetic algorithm simultaneously explores distant regions of parameter space, eventually converging on a high optimum. One of the novel features inherent to the genetic algorithm is a many-fold simultaneous walk through parameter space. In contrast, a gradient search performs a single walk with one configuration (a point in parameter space) containing the information learned about the system up to some point in time. If a fork appears in the road as the gradient search algorithm walks through parameter space, a decision on whether to go one way or the other is made immediately. Once the decision is made, the other option is completely forgotten. If the chosen path ends, there are no other options.

A genetic algorithm is different in that it tries out many roads at the same time. It achieves a parallelized walk through parameter space by using a population of configurations combined with a pairwise reproductive process. When opportunities for reproduction are positively correlated with performance, the population gradually converges over many generations to an optimal region in parameter space.

The difference between a genetic algorithm for optimization problems and a model of genetics is, perhaps, analogous to the difference between a painting and performance art. Whereas a painting is a product, the product of performance art is the process itself. The genetic algorithm applied to optimization problems is like a painting in that the only important aspect is the outcome: Is a good optimum found? The method of finding the optimum and whether it corresponds to real biological phenomena is of little significance. On the other hand, an important feature for a model of genetics is the correct incorporation of underlying biological processes. How the modeled system progresses through time is just as interesting as the final endpoint.

10.4.1.1 Genetic Algorithm Details

Suppose we are given an optimization problem with L optimizable parameters. A genetic algorithm is set up with a population of N individuals with each individual being a haploid entity defined by a single strand of L values corresponding to the optimizable parameters (figure 10.6). The system is initialized with individuals having all parameters chosen randomly from the entire range of allowed parameter values. Thus, an individual represents a specific, randomly chosen point in parameter space and the full population represents a set of many such points.

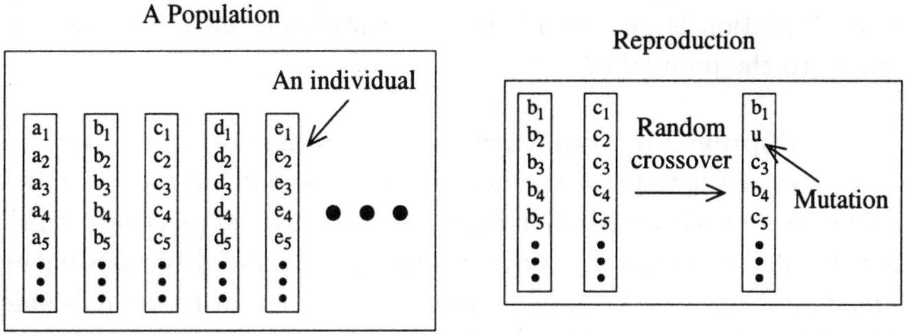

Figure 10.6. Essential genetic algorithm components. A population is composed of many individuals; each individual represents a specific set of parameter values. Reproduction occurs by choosing mates somehow dependent on their fitness values, then applying a crossover algorithm to generate offspring from the two parents. Mutations are also applied to the offspring parameter strands.

Reproduction: Crossover and Mutation Our immediate goal is producing an offspring generation of N individuals. The following description addresses the algorithm used in the simulation code shown below, but alternatives exist (for example see exercise 10.10). First, measure each individual's fitness (performance function value) and determine the average fitness of the population. If an individual's fitness is less than the average, its competitive share of mating opportunities is zero; otherwise, its competitive share is its fitness. Next, loop through each offspring of the next generation. For each offspring, randomly choose two parents weighted according to the competitive share values. Consider two such parents, b and c, each of length L parameters (figure 10.6). Each parameter of the offspring strand is chosen as follows. If a uniform random number is less than the mutation rate μ, a random value is chosen for the offspring's parameter from the allowable range of values, otherwise choose (with equal probability $(1-\mu)/2$) one of the parent values for the offspring strand. Syswerda (1989) called this latter process "uniform crossover," which is not at all like crossing over in diploid organisms.

Both crossover and mutation are important processes. If crossover were the only operation present in the algorithm, we can imagine a situation where one particular parameter value (allele) is observed to perform well at an early generation. Hence, selection favors that one and it may proceed to fixation (the only value for that parameter in the population). Perhaps at a later generation when other parameters have settled onto well-performing values, it turns out that that first parameter's value should be something else. Mutation is one mechanism to continually introduce new parameter values to the population.

Population Structure After reproduction, the offspring population serves as parents of the next generation and the process repeats until convergence is observed. Generally, convergence happens very quickly – the genetic algorithm usually converges too quickly, greedily climbing one of the first major optima it happens upon. The program below implements this type of population structure. Many implementations avoid this greediness by incorporating a population structure that slows down convergence. In natural systems, reproductively isolated populations develop features differing from other populations, even in common environments, due to different founding populations and randomness in the mating process. Allele combinations with a broad range of fitness contributions can dominate very quickly in a small population, reducing genetic diversity and concentrating the population at one local fitness optimum. The time spent stranded at that local optimum depends, in part, on the influx of genetic material that allows exploration of the broader fitness landscape, and that influx depends on having external populations.

Exploiting these biological observations has had positive consequences in genetic algorithm implementations. An interesting method was proposed by Mühlenbein (1989) in which a population was distributed over a two-dimensional lattice with breeding pairs determined in part by spatial proximity. Population structure examined within theoretical population genetics (Felsenstein 1976, Coyne 1992) simply breaks the population into panmictic reproductive groups with an occassional migration of individuals between the different groups. Each group should have a tendency toward a different optimum, but the overall search should not become stalled as easily because of the interactions between groups. This simpler population structure also slows convergence and yields better optima (Wilson, Laidlaw, and Vasudevan 1994).

Schemata Almost any perusal of the genetic algorithm literature will turn up the term *schemata* (Holland 1975). Whether the term schemata contributes conceptual insight, or is a bit of jargon, is open to debate. However, here is my interpretation of schemata. We could, hypothetically, calculate the relative performance for each value of a specific parameter by averaging fitness over all combinations of the other parameters' values. Performing this calculation for each of the parameters would outline the important parameters and their values. In effect, this calculation represents each parameter value's linear contribution to fitness. If there were no interactions between parameters, we would maximize performance by choosing the best value for each parameter. But there are parameter interactions. Hence, we could fix two parameters, each to a specific value, and calculate the relative performance of the pairwise combinations. Extending these calculations to all possible combinations of parameter values would determine all nonlinear contributions to fitness. A single schemata is just one of these possible multiparameter combinations of parameter values; a coarse-grained schemata has only one parameter fixed, whereas the most fine-grained schemata has all parameters fixed (e.g., an individual). We might think of the genetic algorithm as initially discovering well-performing coarse-grained schemata, then gradually moving on to finer- and finer-grained schemata as the algorithm converges on an optimum. Ideally, the best schemata at the finest scale would be the single point in phase space that maximally optimizes the performance function.

Genetic algorithms can be thought of as *schemata processors*. Each individual contains many schemata and represents a single point in the volume of parameter space used to measure each schemata's relative performance. Many of these schemata are contained in subsets of the population's individuals, and each subset serves as an estimate for the respective schemata's relative performance. Reproduction preserves coarse-grained schemata common between two parents (ignoring mutation), but tests them in the new context of the offspring individual. Likewise, new schemata have been generated in each offspring and tested for the first time.

10.4.2 Code

```
#include  <stdlib.h>
#include  <stdio.h>
#include  <time.h>

#define MAXGENS    200
#define NPLANTS    500
#define SEASONLEN  100
#define RRR        0.02
#define MUTATERATE 0.01
#define LZERO      1.0

typedef struct plant_def
{
     double    trait[SEASONLEN];
     double    size;
```

```
    double    fitness;
} Plant;
Plant plants1[NPLANTS], plants2[NPLANTS];

void InitSeed(void);
void InitPlants(Plant *plnt);
void ZeroPlants(Plant *plnt);
void GrowPlants(Plant *plnt, double *ave, double *high);
void ReproducePlants(double avefit,
          Plant *parents, Plant *babies);
void MakeBrakes(double avefit, double *totr,
          double *breaks, Plant *parents);
Plant *GetParent(double totr, double *breaks, Plant *parents);
void Mate(Plant *ma, Plant *pa, Plant *babe);
void PrintPlants(Plant *plnt);

int main(void)
{
     int  ttt;
     double    avefit, highfit;
     Plant     *babes, *adults, *tmpy;

     adults = plants2;
     babes = plants1;
     InitSeed();
     InitPlants(babes);

     for(ttt=0;ttt<MAXGENS;ttt++) /* measurement period */
     {
          tmpy = adults;
          adults = babes;
          babes = tmpy;

          ZeroPlants(adults);
          GrowPlants(adults,&avefit,&highfit);
          ReproducePlants(avefit,adults,babes);
          if(ttt%5==0)
               printf("%d %f %f\n",ttt,avefit,highfit);
     }
     PrintPlants(adults);
     return (0);
}

void InitPlants(Plant *plnt)
{
     int i, j;

     for(i=0;i<NPLANTS;i++,plnt++)
          for(j=0;j<SEASONLEN;j++)
               plnt->trait[j] = drand48();
```

```
}

void ZeroPlants(Plant *plnt)
{
     int i;

     for(i=0;i<NPLANTS;i++,plnt++)
     {
          plnt->size = LZERO;
          plnt->fitness = 0.0;
     }
}

void GrowPlants(Plant *plnt, double *ave, double *high)
{
     int i, j;
     double growth, fitness=0.0;

     *high=0.0;
     for(i=0;i<NPLANTS;i++,plnt++)
     {
          for(j=0;j<SEASONLEN;j++)
          {
               growth = RRR*plnt->size;
               plnt->size += (1.0-plnt->trait[j])*growth;
               plnt->fitness += plnt->trait[j]*growth;
          }
          fitness += plnt->fitness;
          if(plnt->fitness>*high) *high = plnt->fitness;
     }
     *ave = fitness/(double)NPLANTS;
}

void ReproducePlants(double avefit,
          Plant *parents, Plant *babies)
{
     int  i;
     double    totrep;
     double    brakes[NPLANTS];
     Plant     *mom, *dad;

/* measure the total reproduction in the habitat */
     MakeBrakes(avefit, &totrep, brakes, parents);

/* Reproduction */
     for(i=0;i<NPLANTS;i++,babies++)
     {
1.a)      mom = GetParent(totrep, brakes, parents);
          dad = GetParent(totrep, brakes, parents);
          Mate(mom, dad, babies);
```

```
        }
    }

    void MakeBrakes(double avefit, double *totr,
            double *breaks, Plant *parents)
    {
        int i;

        *totr = 0.0;
        for(i=0;i<NPLANTS;i++,parents++)
        {
1.b)        if(parents->fitness>avefit)
                *totr += parents->fitness;
            breaks[i] = *totr;
        }
    }

    Plant *GetParent(double totr, double *breaks, Plant *parents)
    {
        int j, rnum;

        rnum =  drand48() * totr;
        j = (int)((rnum*NPLANTS)/totr);

        if(rnum<breaks[0]) j = 0;
        else if(breaks[j]<rnum)
            while(rnum > breaks[j]) j++;
        else
        {
            while(rnum < breaks[j]) j--;
            j++;
        }
        return(&parents[j]);
    }

    void Mate(Plant *ma, Plant *pa, Plant *babe)
    {
        int j;
        double *mapnt, *papnt, *bapnt, xxx;

        mapnt = ma->trait;
        papnt = pa->trait;
        bapnt = babe->trait;
1.c)    for(j=0;j<SEASONLEN;j++,mapnt++,papnt++,bapnt++)
        {
2.a)        xxx = drand48();
2.b)        if(xxx<MUTATERATE)
2.c)            *bapnt = xxx/MUTATERATE;
2.d)        else if(xxx<0.5+MUTATERATE/2.0)
                *bapnt = *mapnt;
```

```
            else *bapnt = *papnt;
        }
}

void PrintPlants(Plant *plnt)
{
        int i, j, biggie=0;
        double hifit=0.0, avetrait[NPLANTS];

        for(j=0;j<SEASONLEN;j++) avetrait[j] = 0.0;
        for(i=0;i<NPLANTS;i++)
        {
            if(plnt[i].fitness>hifit)
            {
                biggie = i;
                hifit = plnt[i].fitness;
            }
            for(j=0;j<SEASONLEN;j++)
                avetrait[j] += plnt[i].trait[j];
        }
        for(j=0;j<SEASONLEN;j++)
            printf("%d %f %f\n",j,
                avetrait[j]/(double)NPLANTS,
                plnt[biggie].trait[j]);
}
```

10.4.3 Code Details

Line 1: I divided the `ReproducePlants()` routine from the predefined allocation code into a bunch of smaller routines, primarily because at line (1.a) two parents are chosen for each of the babies in the next generation. There are a few minor changes, for example, at line (1.b) within `MakeBrakes()` I only allow those with a fitness above the population average to contribute to the next generation (see exercise 10.2). Choosing mates weighted according to each individual's performance (fitness) is one of the key aspects of the genetic algorithm. Line (1.c) demonstrates combining two parents' traits to obtain an offspring. There are three cases for the allele at a single offspring locus: It is a new mutant value, it comes from mom, or it comes from dad.

Line 2: Here I do something dangerous – use a single random number for multiple purposes. Reusing random numbers can introduce detrimental correlations between various processes, just like a poor random number generator (section 4.2), but it can be done safely. The random number, **xxx**,

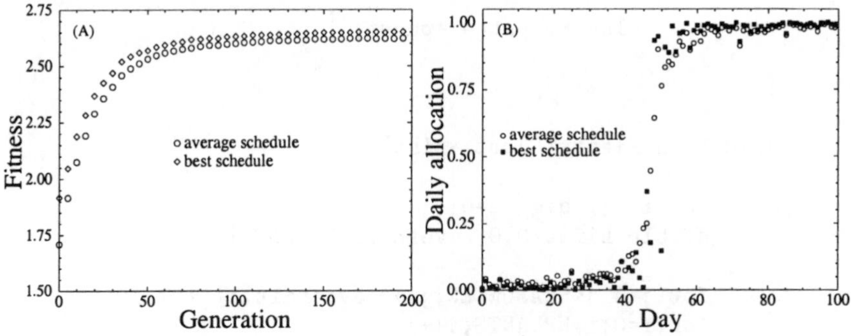

Figure 10.7. (A) Average population fitness and maximum individual fitness, as a function of the generation, reach asymptotic values. (B) Resultant schedules closely match the expected bang–bang schedule.

is drawn at line (2.a). I use it first to check whether a mutation occurs at line (2.b). If mutation occurs, I know 0 <`xxx`<`MUTATERATE`. Thus, at line (2.c), dividing `xxx` by `MUTATERATE` gives random number between 0 and 1 used as the new trait value. If instead there was no mutation at line (2.b), `xxx` is between `MUTATERATE` and 1. I divide this interval into two halves, and if `xxx` is in the first half then I choose the trait value from mom, otherwise from dad. In either event, one random number is used twice. Reusing random numbers only works here because I carefully rescaled my second test.

10.4.4 Simulation Results

The change in average individual fitness over time, figure 10.7A, shows an early, rapid increase, later reaching an asymptotic value. All individuals are initialized with random allocations on each day, so on average, individuals have a daily allocation of 0.5. We might expect, then, that the population's initial fitness corresponds to the fitness of a daily allocation of 0.5 in the first predefined schedule of the first section. In figure 10.2A, the fitness at 0.5 is roughly 1.7 or so, closely matching the initial fitness in figure 10.7A. Within 50 generations the fitness reaches values near the asymptotic fitness.

Figure 10.7B shows the average (open circles) and best (filled squares) allocation schedules at the very end of the simulation. As derived in section 10.3.2, the switch time for both measures is centered right on day 50, the expected day for the parameter values used in the simulation.

Although the genetic algorithm does a good job of finding the optimal allocation schedule, this problem might not be the best test of its abilities. After all, the problem is simple enough to solve analytically and perhaps simpler algorithms such as the gradient search method might work more quickly and precisely. Genetic algorithms really shine when the number of parameters is large and the influence of each parameter on other parameters' optimal values represents a greatly entangled network – the kind of problems that other methods fail to solve.

10.5 Extensions for Stochastic Environments

Now that we compared the simulation of reproducing plants against the analytically predicted bang–bang results, let's extend the code to include several more features. The main process included is herbivory (the removal of photosynthetic biomass by a consumer), since that is an important aspect of a plant's life. We also explore the effects of spatial and temporal stochasticity on the optimal allocation schedules. All of the following runs use modifications of the simulation program described above.

Deterministic Herbivory As a further control against which to compare the more complicated simulations below, suppose that in the one above we added herbivory between days 30 and 40: On each herbivory day the plant loses 5% of its vegetative biomass (after having produced its daily energy and allocating the appropriate fraction to growth). The code needs very few changes, just a line or two within the `GrowPlants()` function,

```
if(j>=herbbegin && j<=herbend) plnt->size *= (1.0-HERBRATE);
```

where `herbbegin` and `herbend` are the first and last days of herbivory, respectively. Results are shown in figure 10.8B, compared with the case of no herbivory (replotted in figure 10.8A). As King and Roughgarden (1982) demonstrated analytically, this form of deterministic herbivory can produce two reproductive peaks, one at the season's end and another just before herbivory hits. The effects on plant size are strong, with size reduced by almost a factor of three in the case of herbivory.

Individual Complexity Next we examine the effects of making life uncertain both in time and space. First, I modified the simulation such that the array of plants represents a one-dimensional habitat (with periodic boundary conditions). Each plant is assured of reproducing one offspring through the maternal route, but matings through the paternal side are roughly proportional to fitness. Each mother takes a small sample of the plants within a mating neighborhood, say of size 10 sites, and chooses the highest fitness mate it finds. The resultant offspring is placed in a randomly chosen site. I also make a plant more complex by including two allocation schedules, schedule 1 defining a plant's allocation schedule if it experiences no herbivory, and schedule 2 being the one it switches to if it experiences herbivory. (This latter extension adds a second `trait` array to the `PLANT` structure.) Further, we will consider that a plant might or might not successfully predict its future herbivory, perhaps using some environmental cue.

As a check of the extended code including two allocation schedules, suppose all plants are attacked by herbivores making a cue of future herbivory irrelevant. A plant starts out using schedule 1, and when hit by herbivores, it immediately switches to schedule 2. The resultant schedule is plotted in figure 10.8C and gives the same results as figure 10.8B. Here the first part of schedule 1 and the last part of schedule 2 experience selection and constitute a net schedule identical to that in figure 10.8B, whereas the other portions of both schedules are randomized about 0.5 because they never experience selection.

Stochastic Herbivory Hits Suppose we introduce stochasticity by having individual plants hit by herbivory with an independent 50% probability. Figure 10.8D shows the resulting allocation schedules for plants that have no cue of future herbivory. The schedules are graded, having lost much of the abruptness associated with the bang–bang schedule; the reproductive allocation slowly increases with the plant's age. One might think that a cue enabling a plant to predict future herbivory would recover both schedules shown in figures 10.8A and B for no herbivory and complete herbivory. Resulting schedules for plants having a 100% accurate cue of when they will be hit with herbivory are shown in figure 10.9A. Both schedules essentially look like those obtained without an herbivory cue, although there may be a bit of an increase in schedule 2 just before herbivory hits. Why doesn't an herbivory cue help? The selection process has problems because plants hit by herbivory usually mate with plants that are not hit (having greater fitness), thus making it more difficult for the response to herbivory selection to be incorporated within the herbivory allocation schedule. Likewise, even though plants that escape herbivory usually mate with other like plants, only half the population avoids herbivory. Perhaps this smaller population size makes determination of the bang–bang schedule more difficult (see exercise 10.5).

Spatially Correlated Stochastic Herbivory Now suppose that instead of random hits of herbivory, herbivores come down and hit big blocks of plants, say randomly selected blocks of 30 plants, much larger than the

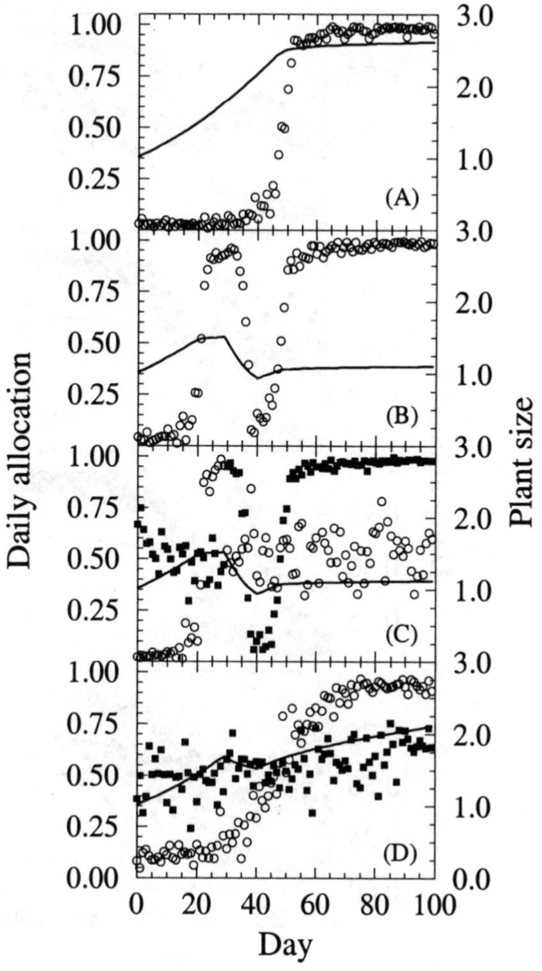

Figure 10.8. Effects of herbivory on plant allocation schedules (symbols) and resultant plant size (curves). (A) Herbivory-free case produces the bang–bang allocation schedule. (B) Deterministic herbivory produces a double flowering period. (C) Deterministic herbivory with pre- and post-herbivory allocation schedules reproduces. (D) Stochasticity in herbivory produces rather graded allocation schedules.

ten-plant mating blocks. Without any cue to herbivory, both schedules show features of the deterministic environments (figure 10.9B). Schedule 1 shows a little blip of reproductive output just before herbivory, and, once hit, schedule 2 produces reproductive material. Both schedules show high reproductive allocation at the season's end. The two schedules become much

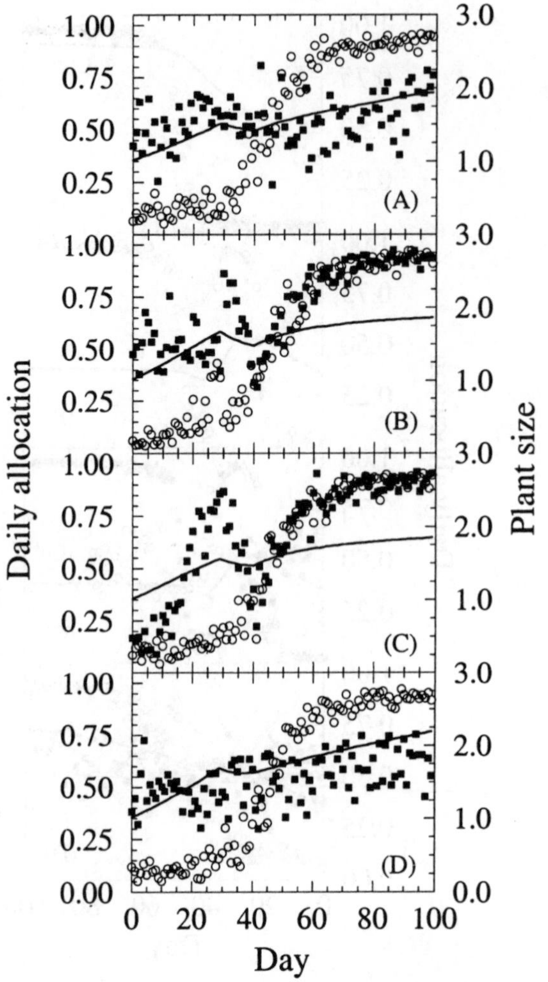

Figure 10.9. Effects of cues to herbivory on plant allocation schedules (symbols) and resultant plant size (curves). (A) Stochastic herbivory, but plants have a perfect cue to their fate. (B) Spatially correlated herbivory with local mating and no herbivory cue. (C) Spatially correlated herbivory with local mating and a perfect herbivory cue. (D) All plants are attacked on day 30, but the last day of herbivory is randomly chosen up to day 50.

cleaner if there is a perfect cue (figure 10.9C), but the schedules are definitely not as clean as in the deterministic cases (figures 10.8A and B). Why is there such a big difference between this case having spatially correlated herbivory hits and the case with spatially random hits? When the two parents experience different environments, the traits benefitting one parent in one environment are crossed over with the same traits that had no influence on the other parent's fitness. In that case, good traits are mixed with irrelevant traits, resulting in poor selection. When herbivory is correlated over spatial scales that are larger than the mating distance, mates experience similar environments, and selection can act on the same set of traits. Hence, good allocation schedules emerge.

Stochastic Herbivory Endpoint As a final example, let's add a variable herbivory period. Suppose all plants get hit starting on day 30 (so no cueing is needed), but the herbivory end date is distributed from day 30 (no herbivory) to day 50. Thus, on average, the plant population experiences the same degree of herbivory as in deterministic herbivory, figures 10.8B and C. Results are shown in figure 10.9D. Again the schedules are very random, with only a general, graded allocation of resource to reproduction. Optimization in the face of stochasticity is difficult.

10.6 Exercises

10.1 Analytically determine the optimal value for c in equation (10.4) and t_s in equation (10.6).

10.2 In the simulation using the genetic algorithm, only the individuals at or above the average fitness can reproduce offspring. What happens if this is changed to higher or lower values? How is the convergence time affected if all plants mate proportionally to their fitness?

10.3 Let herbivores attack and remove a fraction, h, of a plant's vegetative biomass throughout the entire season for the two predefined allocation strategies. Calculate the optimal allocation strategies. How does herbivory affect their relative fitnesses?

10.4 Modify the first extension of the GA simulation (using plants with only one allocation schedule) to include herbivory on the reproductive biomass coincident with herbivory on the vegetative biomass. Does the dual flowering period remain?

10.5 Explore why the GA with herbivory-cued plants doesn't produce the clean allocation strategies found for the deterministic cases. Is it an effectively reduced population size? Compare the bang–bang strategies obtained for a variety of population sizes. Is it crossing-over of traits not undergoing selection? Fix the population size and for some fraction of plants choose mates randomly rather than based on fitness. Compare the resulting schedules for several values of this fraction.

10.6 Refer to King and Roughgarden's (1982) work and incorporate herbivory within the simulation in such a way that an analytic match can be made.

10.7 Reproduce the results of the Spatially Correlated Stochastic Herbivory portion of section 10.5 after extending the GA simulation as described.

10.8 Simulate plants in a desert facing two rainy periods, spring and fall, with a rain-free (and no-growth) period in the summer. Suppose that during days 30 to 40 the growth rate is zero. Determine and explain the optimal schedule (see King and Roughgarden 1982).

10.9 (a) Implement a gradient search method to locate the optimum of equation (10.7). (b) Extend the program to solve for the optimal daily allocation schedule.

10.10 There are other ways to impose selection on the genetic algorithm's population. For example, given a fixed population number, competition for the limited berths removes poorly performing individuals. Modify the simulation code in the following way (Wilson, Laidlaw, and Vasudevan 1994): 1) Rank all individuals according to fitness with the highest fitness individual as #1 and the lowest as #N. 2) With probability one-half mate #1 and #2, #3 and #4, etc.; alternatively mate #2 and #3, #4 and #5, etc., with #1 and #N passing unchanged to the next generation. 3) Produce two offspring from each mating pair and measure the offsprings' fitnesses. Pass the best two individuals of the four individuals of each "family" (two parents and two offspring) to the next generation. How does this modification affect convergence?

11

Epilogue

Simulation and mathematical approaches both have their advantages and disadvantages. The clarity of biological assumptions contained in a mathematical formalization is extremely appealing, but mathematical tractability is often constraining.[1] The multitude of simulation assumptions can be frustrating, but the questions we can ask using a simulation approach are almost limitless. It is my firm belief that the interplay between mathematical and simulation models helps advance theoretical ecology and evolution, much like the interplay between theory and empirical studies helps advance the study of ecology and evolution (figure 11.1).

The interplay between the two approaches is especially crucial for simulation work. As I demonstrated in chapter 2, the ordering of a few lines of code can strongly influence the results, at least in a quantitative sense. Precisely because such programming details are important, yet their implications are so difficult to assess and communicate scientifically, simulations alone will never serve as the language for the foundation of theoretical ecology.

Mathematical theorists can begin with a precisely formulated mathematical problem having clearly stated biological assumptions. The starting point is explicit, and there is only one correct solution emanating from it, even though there are many alternative approximations to it. A mathematical approach can therefore proceed without appeal to simulation results; indeed, a simulation introduces stochastic features that the mathematical framework may have disregarded entirely. Hence, the simulation would be completely irrelevant to the solution of the *mathematical* problem. But I argue that the really important issue is whether the mathematical formulation addresses the biological question.

Although the starting point for a simulation model seems relatively vague,

[1] Numerical solutions greatly relax this constraint.

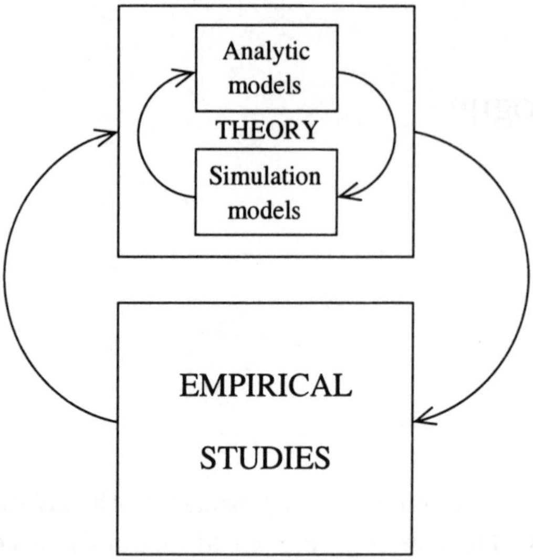

Figure 11.1. Ideal interplay between simulation and mathematical models parallels the interplay between theory and empirical studies.

it is only because the incorporation of individual-based, spatially explicit interactions reveals hidden assumptions. Mathematical models are limited in their tractability in the face of discrete, stochastic, age-, size-, and state-structured individuals, yet these are the conditions of the real biological world. More robust theoretical predictions can arise by examining mathematically derived ecological and evolutionary predictions under the light of simulation models. If, through simulation work, a previously unrecognized feature is observed to have a qualitative influence on ecological or evolutionary dynamics, then that feature must be addressed by theorists through a mathematical approach before asking empiricists to devise new experimental inquiries.

I have shown several examples where the interplay can work. Individual-based simulations point out the dynamical importance of a population being packaged in the discrete unit of a stochastically interacting individual. A dramatic demonstration was found in the predator–prey simulations of chapter 9 where the size of local populations was manipulated (figure 9.11 on page 245) with profound influence on resultant global dynamics. It is only when local populations are large that the simulation model may not add any new insight beyond that provided by the analysis of deterministic mathematical models incorporating population densities (Wilson 1998). Spatial effects in these simulations can be very important. Qualitative differences were observed between the predictions of nonspatial and spatial models for female–hermaphrodite coexistence in chapter 8. The nonspatial model predicted the collapse of the gynodioecious mating system, but the spatial simulation demonstrated coexistence with a spatiotemporally dynamic population distribution. We also observed animal grouping arising out of a simple foraging model (figure 7.11 on page 179), as well as the spatiotemporal dynamical richness of reaction–diffusion models (chapter 9).

Benchmarking simulations against analytic models can have a more mundane purpose than the loftier goal of advancing theoretical insight. Comparing simulation results to those of an analytic model, at least in limiting cases, builds confidence that the program was written as intended. It is a very useful process: For the immigration–emigration model in chapter 2, figure 2.4, there is some disagreement between analytic and simulation results, which was greatly resolved in figure 2.6 upon discerning a subtlety in the simulation model.

When comparing results, it may be unclear when close is close enough. The simplistic foraging model of section 7.1 in chapter 7, in spite of its crude assumptions, did a fantastic job characterizing the simulated forager's average intake rate (figure 7.7) The pursuit of theoretical parsimony demands that we stop with that model! It captured the essence of the mechanisms behind the problem very well, and that essence was only diluted by the more complicated size-structured analytic model that only went part way in resolving the differences (figure 7.10). Do we conclude that any remaining disagreement arises from programming errors or the true effects of a mathematically inaccessible biological feature? The simplistic model's predictions might break down for more detailed questions involving variances in rates and resource abundance, necessitating more complicated analytic descriptions. When do we stop writing successively detailed analytic models? The process of resolving these differences is surely analogous to the interpretation of empirical results (figure 11.1). We stop adding details to a model and accept imprecision when we feel a mechanistic explanation is sufficiently complete.

C is an ideal computer programming language with which to carry out this scientific work. Its flexibility and structure are outstanding. Certainly, many other computing languages work equally well; some of them are higher-level languages compiled using software written in C, allowing complete C programs within. C will be around for many years to come: The connection between C and the Unix and open-source Linux operating systems, with millions of users world-wide, ensures the language's survival. I hope that you have gained a broad foundation of programs and routines from which to base further explorations.

References

Abrams, P.A. 1982. Functional responses of optimal foragers. *American Naturalist* **120**: 382–390.

Abundo, M. 1991. A stochastic model for predator–prey systems: basic properties, stability and computer simulation. *Journal of Mathematical Biology* **29**: 495–511.

Alexander, R.McN. 1996. *Optima for animals.* Princeton University Press, Princeton.

Allee, W.C. 1938. *The social life of animals.* Norton and Co., New York.

Allen, E.J., L.J.S. Allen, and X. Gilliam. 1996. Dispersal and competition models for plants. *Journal of Mathematical Biology* **34**: 455–481.

Atkinson, K.E. 1978. *An introduction to numerical analysis.* John Wiley, New York.

Atkinson, W.D. and B. Shorrocks. 1981. Competition on a divided and ephemeral resource: a simulation model. *Journal of Animal Ecology* **50**: 461–471.

Banks, R.B. 1994. *Growth and diffusion phenomena: mathematical frameworks and applications.* Springer-Verlag, Berlin.

Bartlett, M.S. 1955. *An introduction to stochastic processes with special reference to methods and applications.* Cambridge University Press, Cambridge.

Bartlett, M.S. 1956. On theoretical models for competitive and predatory biological systems. *Biometrika* **44**: 27–42.

Bartlett, M.S. 1960. *Stochastic population models.* John Wiley, London.

Bascompte, J. and R.V. Solé. 1995. Rethinking complexity: modelling spatiotemporal dynamics in ecology. *Trends in Ecology and Evolution* **10**: 361–366.

Bazzaz, F.A., N.R. Chiariello, P.D. Coley, and L.F. Pitelka. 1987. Allocating resources to reproduction and defense. *BioScience* **37**: 58–67.

Bolker, B.M. and S.W. Pacala. 1999. Spatial moment equations for plant competition: understanding spatial strategies and the advantages of short dispersal. *American Naturalist* **153**: 575–602.

Boon, J.P., D. Dab, R. Kapral, and A. Lawniczak. 1996. Lattice gas automata for reactive systems. *Physics Reports* **273**: 55–147.

Bounds, D.G. 1987. New optimization methods from physics and biology. *Nature* **329**: 215–219.

Box, G.E.P. and M.E. Muller. 1958. A note on the generation of random normal deviates. *The Annals of Mathematical Statistics* **29**: 610-611.

Brown, D.B. and R.I.C. Hansell. 1987. Convergence to an evolutionarily stable strategy in the two-policy game. *American Naturalist* **130**: 929–940.

Bulmer, M. 1994. *Theoretical evolutionary ecology.* Sinauer Associates, Sunderland.

Burnett, T. 1958. A model of host–parasite interaction. *Proceedings Tenth International Congress of Entomology* **2**: 679–686.

Case, T.J. 2000. *An illustrated guide to theoretical ecology.* Oxford University Press, New York.

Caswell, H. 1972. A simulation study of a time lag population model. *Journal of Theoretical Biology* **34**: 419–439.

Caswell, H. 1988. Theory and models in ecology: a different perspective. *Ecological Modelling* **43**: 33–44.

Charlesworth, B. and D. Charlesworth. 1978. A model for the evolution of dioecy and gynodioecy. *American Naturalist* **112**: 975–997.

Charnov, E.L. 1976. Optimal foraging, the marginal value theorem. *Theoretical Population Biology* **9**: 129–136.

Charnov, E.L. 1982. *The theory of sex allocation.* Princeton University Press, Princeton.

Chesson, P.L. 1981. Models for spatially distributed populations: the effect of within-patch variability. *Theoretical Population Biology* **19**: 288–325.

Clark, J.S. 1998. Why trees migrate so fast: confronting theory with dispersal biology and the paleorecord. *American Naturalist* **152**: 204–224.

Clark, J.S., M. Silman, R. Kern, E. Macklin, and J. HilleRisLambers. 1999. Seed dispersal near and far: patterns across temperate and tropical forests. *Ecology* **80**: 1475–1494.

Clément, E., L.M. Sander, and R. Kopelman. 1989. Source-term and excluded-volume effects on the diffusion-controlled a+b→0 reaction in one dimension: rate laws and particle distributions. *Physical Review* **A 39**: 6455–6465.

Cohen, D. 1967. Optimizing reproduction in a randomly varying environment when a correlation may exist between the conditions at the time a choice has to be made and the subsequent outcome. *Journal of Theoretical Biology* **16**: 1–14.

Cohen, D. and H. Parnas. 1976. An optimal policy for the metabolism of storage materials in unicellular algae. *Journal of Theoretical Biology* **56**: 1–18.

Collier, T.R., W.W. Murdoch, and R.M. Nisbet. 1994. Egg load and the decision to host-feed in the parasitoid, Aphytis melinus. *Journal of Animal Ecology* **63**: 299–306.

Comins, H.N. and M.P. Hassell. 1987. The dynamics of predation and competition in patchy environments. *Theoretical Population Biology* **31**: 393–421.

Comins, H.N., M.P. Hassell, and R.M. May. 1992. The spatial dynamics of host–parasitoid systems. *Journal of Animal Ecology* **61**: 735–748.

Coppersmith, S.N., C.-H. Liu, S. Majumdar, O. Narayan, and T.A. Witten. 1995. A model for force fluctuations in bead packs. *Physical Review* **E 53**: 4673–4685.

Costantino, R.F., J.M. Cushing, B. Dennis, and R.A. Desharnais. 1995. Experimentally induced transitions in the dynamic behaviour of insect populations. *Nature* **375**: 227–230.

Coyne, J.A. 1992. Genetics and speciation. *Nature* **355**: 511–515.

Crowley, P.H. 1981. Dispersal and the stability of predator–prey interactions. *American Naturalist* **118**: 673–701.

Davies, N.B. and A.I Houston. 1981. Owners and satellites: the economics of territory defence in the pied wagtail, *Motacilla alba*. *Journal of Animal Ecology* **50**: 157–180.

DeAngelis, D.L. 1976. Application of stochastic models to a wildlife population. *Mathematical Biosciences* **31**: 227-236.

de Roos, A.M. 1996. A gentle introduction to physiologically structured population models. In S. Tuljapurkar and H. Caswell, eds., *Structured-population models in marine, terrestrial, and freshwater systems*. Chapman and Hall, New York.

de Roos, A.M., E. McCauley, and W.G. Wilson. 1991. Mobility versus density-limited predator–prey dynamics on different spatial scales. *Proceedings of the Royal Society of London* **B 246**: 117–122.

de Roos, A.M., O. Diekmann, and J.A.J. Metz. 1992. Studying the dynamics of structured population models: a versatile technique and its application to daphnia. *American Naturalist* **139**: 123–147.

Doi, M. 1976. Stochastic theory of diffusion-controlled reaction. *Journal of Physics* **A 9**: 1479–1495.

Donalson, D.D. and R.M. Nisbet. 1999. Population dynamics and spatial scale: effects of system size on population persistence. *Ecology* **80**: 2492-2507.

Durrett, R. and S. Levin. 1994. The importance of being discrete (and spatial). *Theoretical Population Biology* **46**: 363–394.

Durrett, R. and S. Levin. 1997. Allelopathy in spatially distributed populations. *Journal of Theoretical Biology* **185**: 165–171.

Durrett, R. and S. Levin. 1998. Spatial aspects of interspecific competition. *Theoretical Population Biology* **53**: 30–43.

Dwyer, G. 1992. On the spatial spread of insect pathogens: theory and experiment. *Ecology* **73**: 479–494.

Ellner, S.P., A. Sasaki, Y. Haraguchi, and H. Matsuda. 1998. Speed of invasion in lattice population models: pair-edge approximation. *Journal of Mathematical Biology* **36**: 469–484.

Engen, S. and B.-E. Saether. 1994. Optimal allocation of resources to growth and reproduction. *Theoretical Population Biology* **46**: 232–248.

Felsenstein, J. 1976. The theoretical population genetics of variable selection and migration. *Annual Review of Genetics* **10**: 253–280.

Frank, S.A. 1989. The evolutionary dynamics of cytoplasmic male sterility. *American Naturalist* **131**: 345–376.

Gilpin, M.E. 1973. Do hares eat lynx? *American Naturalist* **107**: 727–730.

Goel, N.S. and N. Richter-Dyn. 1974. *Stochastic models in biology*. Academic Press, New York.

Goldberg, D.E. 1989. *Genetic algorithms in search, optimization, and machine learning*. Addison-Wesley, Reading.

Gottfried, B. 1996. *Programming with C, 2nd edition. Schaum's Outline Series*. McGraw-Hill, New York.

Gouyon, P.H., F. Vichot, and J.M.M. VanDamme. 1991. Nuclear–cytoplasmic male sterility: single-point equilibria versus limit cycles. *American Naturalist* **137**: 498–514.

Grimm, V. 1999. Ten years of individual-based modelling in ecology: what have we learned and what could we learn in the future? *Ecological Modelling* **115**: 129–148.

Gueron, S. and N. Liron. 1989. A model of herd grazing as a travelling wave, chemotaxis and stability. *Journal of Mathematical Biology* **27**: 595–608.

Gurney, W.S.C., S.P. Blythe, and R.M. Nisbet. 1980. Nicholson's blowflies revisited. *Nature* **287**: 17–21.

Gurney, W.S.C. and R.M. Nisbet. 1976. Spatial pattern and the mechanism of population regulation. *Journal of Theoretical Biology* **59**: 361–370.

Gurney, W.S.C. and R.M. Nisbet. 1983. The systematic formulation of delay-differential models of age or size structured populations. In H.I. Freedman, and C. Strobeck, eds., *Population Biology.* Springer-Verlag, New York.

Gurney, W.S.C. and R.M. Nisbet. 1998. *Ecological Dynamics.* Oxford University Press, New York.

Hadeler, K.P., U. an der Heiden, and F. Rothe. 1974. Nonhomogeneous spatial distributions of populations. *Journal of Mathematical Biology* **1**: 165–176.

Hanski, I. and D.Y. Zhang. 1993. Migration, metapopulation dynamics and fugitive coexistence. *Journal of Theoretical Biology* **163**: 491–504.

Harada, Y., H. Ezoe, Y. Iwasa, H. Matsuda, and K. Sato. 1995. Population persistence and spatially limited social interaction. *Theoretical Population Biology* **48**: 65–91.

Hart, D.R. and R.H. Gardner. 1997. A spatial model for the spread of invading organisms subject to competition. *Journal of Mathematical Biology* **35**: 935–948.

Hastings, A. 1980. Disturbance, coexistence, history and competition for space. *Theoretical Population Biology* **18**: 363–373.

Hastings, A. 1997. *Population biology: concepts and models.* Springer-Verlag, New York.

Hilborn, R. 1975. The effect of spatial heterogeneity on the persistence of predator–prey interactions. *Theoretical Population Biology* **8**: 346–355.

Hogeweg, P. 1988. Cellular automata as a paradigm for ecological modeling. *Applied Mathematics and Computation* **27**: 81–100.

Holgate, P. 1967. The size of elephant herds. *The Mathematical Gazette* **51**: 302–304.

Holland, J.H. 1975. *Adaptation in natural and artificial systems.* University of Michigan Press, Ann Arbor.

Holmes, E.E., M.A. Lewis, J.E. Banks, and R.R. Veit. 1994. Partial differential equations in ecology: spatial interactions and population dynamics. *Ecology* **75**: 17–29.

Holmes, E.E. and H.B. Wilson. 1998. Running from trouble: long-distance dispersal and the competitive coexistence of inferior species. *American Naturalist* **151**: 578–586.

Holyoak, M. and S.P. Lawler. 1996. Persistence of an extinction-prone predator–prey interaction through metapopulation dynamics. *Ecology* **77**: 1867–1879.

Huffaker, C.B. 1958. Experimental studies on predation: dispersion factors and predator–prey oscillations. *Hilgardia* **27**: 343–383.

Huston, M., D. DeAngelis, and W. Post. 1988. New computer models unify ecological theory. *BioScience* **38**: 682–691.

Intriligator, M.D. 1971. *Mathematical optimization and economic theory.* Prentice-Hall, Englewood Cliffs, NJ.

Ives, A.R. 1988. Covariance, coexistence and the population dynamics of two competitors using a patchy resource. *Journal of Theoretical Biology* **133**: 345–361.

Iwasa, Y. 1989. Pessimistic plant: optimal growth schedule in stochastic environments. *Theoretical Population Biology* **40**: 246–268.

Iwasa, Y. and D. Cohen. 1989. Optimal growth schedule of a perennial plant. *American Naturalist* **133**: 480–505.

Iwasa, Y. and J. Roughgarden. 1984. Shoot/root balance of plants: optimal growth of a system with many vegetative organs. *Theoretical Population Biology* **25**: 78–105.

Kang, K. and S. Redner. 1985. Fluctuation-dominated kinetics in diffusion-controlled reactions. *Physical Review* **A 32**: 435–447.

Kareiva, P. 1982. Experimental and mathematical analyses of herbivore movement: quantifying the influence of plant spacing and quality on foraging discrimination. *Ecological Monographs* **52**: 261–282.

Karlin, S. and H.M. Taylor. 1975. *A first course in stochastic processes*, 2nd edition. Academic Press, San Diego.

Karlin, S. and H.M. Taylor. 1981. *A second course in stochastic processes.* Academic Press, San Diego.

Kehr, K.W., R. Kutner, and K. Binder. 1981. Diffusion in concentrated lattice gases. self-diffusion of noninteracting particles in three-dimensional lattices. *Physical Review* **B 23**: 4931–4945.

Keitt, T.H. and A.R. Johnson. 1995. Spatial heterogeneity and anomalous kinetics: emergent patterns in diffusion-limited predatory–prey interaction. *Journal of Theoretical Biology* **172**: 127–139.

Keller, E.F. and L.A. Segel. 1971. Traveling bands of chemotactic bacteria: a theoretical analysis. *Journal of Theoretical Biology* **30**: 235–248.

Kelley, A. and I. Pohl. 1998. *A book on C*, 4th edition. Addison Wesley Longman, Reading.

Kernighan, B.W. and D.M. Ritchie. 1988. *The C programming language*, 2nd edition. Prentice Hall, Englewood Cliffs, NJ.

King, D. and J. Roughgarden. 1982. Multiple switches between vegetative and reproductive growth in annual plants. *Theoretical Population Biology* **21**: 194–204.

Kirkpatrick, S., C.D. Gelatt, and M.P. Vecchi. 1983. Optimization by simulated annealing. *Science* **220**: 671–680.

Kirkpatrick, S. and E.P. Stoll. 1981. A very fast shift-register sequence random number generator. *Journal of Computational Physics* **40**: 517–526.

Kishimoto, K. 1990. Coexistence of any number of species in the Lotka–Volterra competitive system over two-patches. *Theoretical Population Biology* **38**: 149–158.

Kooijman, S.A.L.M. 1993. *Dynamic energy budgets in biological systems.* Cambridge University Press, Cambridge.

Kopelman, R. 1986. Rate processes on fractals: theory, simulations and experiments. *Journal of Statistical Physics* **42**: 185–200.

Kot, M. and W.M. Schaffer. 1986. Discrete-time growth–dispersal models. *Mathematical Biosciences* **80**: 109–136.

Krebs, J.R. and A. Kacelnik. 1991. Decision-making. In J.R. Krebs and N.B. Davies, eds., *Behavioral ecology: an evolutionary approach*, Blackwell Scientific Publications, Oxford.

Larralde, H., P. Trunfio, S. Havlin, H.E. Stanley, and G.H. Weiss. 1992. Territory covered by N diffusing particles. *Nature* **355**: 423–426.

Leslie, P.H. 1958. A stochastic model for studying the properties of certain biological systems by numerical methods. *Biometrika* **45**: 16–31.

Leslie, P.H. and J.C. Gower. 1958. The properties of a stochastic model for two competing species. *Biometrika* **45**: 316–330.

Leslie, P.H. and J.C. Gower. 1960. The properties of a stochastic model for the predator–prey type of interaction between two species. *Biometrika* **47**: 219–234.

Levin, S.A., D. Cohen, and A. Hastings. 1984. Dispersal strategies in patchy environments. *Theoretical Population Biology* **26**: 165–191.

Levin, S.A. and L.A. Segel. 1985. Pattern generation in space and aspect. *SIAM Review* **27**: 45–67.

Levins, R. 1969. The effect of random variations of different types on population growth. *Zoology* **62**: 1061–1065.

Lewis, M.A. 1994. Spatial coupling of plant and herbivore dynamics: the contribution of herbivore dispersal to transient and persistent "waves" of damage. *Theoretical Population Biology* **45**: 277–312.

Lloyd, D.G. 1975. The maintenance of gynodioecy and androdioecy in angiosperms. *Genetica* **45**: 325–339.

Lloyd, D.G. and D.L. Venable. 1992. Some properties of natural selection with single and multiple constraints. *Theoretical Population Biology* **41**: 90–110.

Lotka, A.J. 1924. *Elements of physical biology.* Williams and Wilkins, Baltimore.

Luckinbill, L.S. 1973. Coexistence in laboratory populations of *Paramecium aurelia* and its predator *Didinium nasutum*. *Ecology* **54**: 1320–1327.

Mangel, M. and C.W. Clark. 1988. *Dynamic modeling in behavioral ecology.* Princeton University Press, Princeton.

Mangel, M. and C. Tier. 1993. Dynamics of metapopulations with demographic stochasticity and environmental catastrophes. *Theoretical Population Biology* **44**: 1–31.

MacArthur, R. and R. Levins. 1967. The limiting similarity, convergence, and divergence of coexisting species. *American Naturalist* **101**: 377–385.

Matsuda, H., N. Ogita, A. Sasaki, and K. Sato. 1992. Statistical mechanics of population: the lattice Lotka–Volterra model. *Progress of Theoretical Physics* **88**: 1035–1049.

Maynard Smith, J. 1978. Optimization theory in evolution. *Annual Reviews in Ecology and Systematics* **9**: 31–56.

McCauley, D.E. and D.R. Taylor. 1997. Local population structure and sex ratio: evolution in gynodioecious plants. *American Naturalist* **150**: 406–419.

McCauley, E., W.G. Wilson, and A.M. de Roos. 1993. Dynamics of age-structured and spatially structured predator–prey interactions: individual-based models and population-level formulations. *American Naturalist* **142**: 412–442.

McCauley, E., R.M. Nisbet, A.M. De Roos, W.W. Murdoch, and W.S. Gurney. 1996. Structured population models of herbivorous zooplankton. *Ecological Monographs* **66**: 479–501.

McLaughlin, J.F. and J. Roughgarden. 1991. Pattern and stability in predator–prey communities: how diffusion in spatially variable environments affects the Lotka–Volterra model. *Theoretical Population Biology* **40**: 148–172.

Metropolis, N., A.W. Rosenbluth, M.N. Rosenbluth, A.H. Teller, and E. Teller. 1953. Equation of state calculations by fast computing machines. *The Journal of Chemical Physics* **21**: 1087–1092.

Milinski, M. 1979. An evolutionarily stable feeding strategy in sticklebacks. *Zeitschrift für Tierpsychologie* **51**: 36–40.

Mollison, D. 1977. Spatial contact models for ecological and epidemic spread. *Journal of the Royal Statistical Society* **B 39**: 283–326.

Molofsky, J. 1994. Population dynamics and pattern formation in theoretical populations. *Ecology* **75**: 30–39.

Morgan, B.J.T. 1976. Stochastic models of grouping changes. *Advances in Applied Probability* **8**: 30–57.

Morgan, B.J.T. 1984. *Elements of simulation* Chapman and Hall, London.

Morin, P.J. 1999. *Community ecology.* Blackwell Science, Malden.

Mullish, H. and H.L. Cooper. 1987. *The spirit of C: an introduction to modern programming.* West Publishing, St. Paul.

Murray, J.D. 1989. *Mathematical biology.* Springer-Verlag, Berlin.

Neubert, M.G., M. Kot, and M.A. Lewis. 1995. Dispersal and pattern formation in a discrete-time predator–prey model. *Theoretical Population Biology* **48**: 7–43.

Neuhauser, C. 1998. Habitat destruction and competitive coexistence in spatially explicit models with local interactions. *Journal of Theoretical Biology* **193**: 445–463.

Nisbet, R.M., S. Diehl, W.G. Wilson, S.D. Cooper, D.D. Donalson, and K. Kratz. 1997. Primary-productivity gradients and short-term population dynamics in open systems. *Ecological Monographs* **67**: 535–553.

Nisbet, R.M. and Gurney, W.S.C. 1982. *Modelling fluctuating populations.* John Wiley, New York.

Okubo, A. 1980. *Diffusion and ecological problems: mathematical models.* Springer-Verlag, Heidelberg.

Othmer, H.G., S.R. Dunbar, and W. Alt. 1988. Models of dispersal in biological systems. *Journal of Mathematical Biology* **26**: 263–298.

Oualline, S. 1993. *Practical C programming.* O'Reilly & Associates, Sebastopol.

Pacala, S.W. and S.A. Levin. 1997. Biologically generated spatial pattern and the coexistence of competing species. In D. Tilman and P. Kareiva, eds., *Spatial ecology*, Princeton University Press, Princeton.

Pais, A. 1982. *Subtle is the Lord: the science and the life of Albert Einstein.* Oxford University Press, Oxford.

Pannell, J. 1997. The maintenance of gynodioecy and androdioecy in a metapopulation. *Evolution* **51**: 10–20.

Parnas, H. and D. Cohen. 1976. The optimal strategy for the metabolism of reserve materials in microorganisms. *Journal of Theoretical Biology* **56**: 19–55.

Perrin, N. and R.M. Sibly. 1993. Dynamic models of energy allocation and investment. *Annual Reviews in Ecology and Systematics* **24**: 379–410.

Pielou, E.C. 1969. *An introduction to mathematical ecology.* John Wiley, New York.

Poole, R.W. 1974. A discrete time stochastic model of a two prey, one predator species interaction. *Theoretical Population Biology* **5**: 208–228.

Possingham, H.P. 1989. The distribution and abundance of resources encountered by a forager. *American Naturalist* **133**: 42–60.

Possingham, H.P. and A.I. Houston. 1990. Optimal patch use by a territorial forager. *Journal of Theoretical Biology* **145**: 343–353.

Press, W.H., B.P. Flannery, S.A. Teukolsky and W.T. Vetterling. 1986. *Numerical recipes: the art of scientific computing.* Cambridge University Press, Cambridge.

Pugliese, A. 1988. Optimal resource allocation in perennial plants: a continuous time model. *Theoretical Population Biology* **34**: 215–247.

Pyke, G.H. 1983. Animal movements: an optimal foraging approach. In I.R. Swingland and R.J. Greenwood, eds., *The ecology of animal movement*, Oxford University Press, Oxford.

Rand, D.A., M. Keeling, and H.B. Wilson. 1995. Invasion, stability and evolution to criticality in spatially extended, artificial host–pathogen ecologies. *Proceedings of the Royal Society of London* **B 259**: 55–63.

Renshaw, E. 1991. *Modelling biological populations in space and time.* Cambridge University Press, Cambridge.

Richards, S.A. 1997. Completed Richarson extrapolation in space and time. *Communications in Numerical Methods in Engineering* **13**: 573–582.

Roff, D.A. 1974a. Spatial heterogeneity and the persistence of populations. *Oecologia* **15**: 245–258.

Roff, D.A. 1974b. The analysis of a population model demonstrating the importance of dispersal in a heterogeneous environment. *Oecologia* **15**: 259–275.

Roff, D.A. 1983. An allocation model of growth and reproduction in fish. *Canadian Journal of Fisheries and Aquatic Sciences* **40**: 1395–1404.

Rohani, P. and O. Miramontes. 1995. Host–parasitoid metapopulations: the consequences of parasitoid aggression on spatial dynamics and searching efficiency. *Proceedings of the Royal Society of London* **B 260**: 335–342.

Rohani, P., T.J. Lewis, D. Grunbaum, and G.D. Ruxton. 1997. Spatial self-organization in ecology: pretty patterns or robust reality? *Trends in Ecology and Evolution* **12**: 70–74.

Rosen, R. 1967. *Optimality principles in biology.* Butterworths, London.

Ross, A.H. and R.M. Nisbet. 1990. Dynamic models of growth and reproduction of the mussel *Mytilus edulis* L. *Functional Ecology* **4**: 777–787.

Roughgarden, J. 1975. A simple model for population dynamics in stochastic environments. *American Naturalist* **109**: 713–736.

Ruxton, G.D. and P. Rohani. 1996. The consequences of stochasticity for self-organized spatial dynamics, persistence and coexistence in spatially extended host–parasitoid communities. *Proceedings of the Royal Society of London* **B 263**: 625–631.

Satō, K., H. Matsuda, and A. Sasaki. 1994. Pathogen invasion and host extinction in lattice structured populations. *Journal of Mathematical Biology* **32**: 251–268.

Satulovsky, J.E. and T. Tomé. 1997. Spatial instabilities and local oscillations in a lattice gas Lotka–Volterra model. *Journal of Mathematical Biology* **35**: 344–358.

Schaffer, W.M. 1983. The application of optimal control theory to the general life history problem. *American Naturalist* **121**: 418–431.

Schmitz, O.J. and G. Booth. 1997. Modelling food web complexity: the consequences of individual-based, spatially explicit behavioural ecology on trophic interactions. *Evolutionary Ecology* **11**: 379–398.

Schwinning, S. and A.J. Parsons. 1996a. Analysis of the coexistence mechanisms for grasses and legumes in grazing systems. *Journal of Ecology* **84**: 799–813.

Schwinning, S. and A.J. Parsons. 1996b. A spatially explicit population model of stoloniferous N-fixing legumes in mixed pasture with grass. *Journal of Ecology* **84**: 815–826.

Segel, L.A. and J.L. Jackson. 1972. Dissipative structure: an explanation and an ecological example. *Journal of Theoretical Biology* **37**: 545–559.

Shorrocks, B., W. Atkinson, and P. Charlesworth. 1979. Competition on a divided and ephemeral resource. *Journal of Animal Ecology* **48**: 899–908.

Silvertown, J., S. Holtier, J. Johnson, and P. Dale. 1992. Cellular automaton models of interspecific competition for space – the effect of pattern on process. *Journal of Ecology* **80**: 527–534.

Sinko, J.W. and W. Streifer. 1967. A new model for age–size structure of a population. *Ecology* **48**: 910–918.

Skellam, J.G. 1951. Random dispersal in theoretical populations. *Biometrika* **38**: 196–218.

Smith, C.H. 1983. Spatial trends in canadian snowshoe hare, *Lepus americanus*, population cycles. *Canadian Field-Naturalist* **97**: 151–160.

Smith, M. 1991. Using massively parallel supercomputers to model stochastic spatial predator–prey systems. *Ecological Modelling* **58**: 347–367.

Sneppen, K., P. Bak, H. Flyvbjerg, and M.H. Jensen. 1995. Evolution as a self-organized critical phenomenon. *Proceedings of the National Acadamy of Science, USA* **92**: 5209–5213.

Stenseth, C. 1980. Spatial heterogeneity and population stability: some evolutionary consequences. *Oikos* **35**: 165–184.

Stephens, D.W. and J.R. Krebs. 1986. *Foraging theory.* Princeton University Press, Princeton.

Stokes, T.K., W.S.C. Gurney, R.M. Nisbet, and S.P. Blythe. 1988. Parameter evolution in a laboratory insect population. *Theoretical Population Biology* **34**: 248–265.

Strikwerda J.C. 1989. *Finite difference schemes and partial differential equations.* Wadsworth and Brooks/Cole, Pacific Grove.

Sumida, B.H., A.I. Houston, J.M. McNamara, and W.D. Hamilton. 1990. Genetic algorithms and evolution. *Journal of Theoretical Biology* **147**: 59–84.

Syswerda, G. 1989. Uniform crossover in genetic algorithms. In J.D. Schaffer, ed., *Proceedings of the Third International Conference on Genetic Algorithms*, Morgan Kaufmann, San Mateo.

Taylor, A.D. 1992. Deterministic stability analysis can predict the dynamics of some stochastic population models. *Journal of Animal Ecology* **61**: 241–248.

Tilman, D. 1987. The importance of the mechanisms of interspecific competition. *American Naturalist* **129**: 769–774.

Tilman, D. and P. Kareiva, eds., 1997. *Spatial ecology: the role of space in population dynamics and interspecific interactions.* Princeton University Press, Princeton.

Toquenaga, Y. and M.J. Wade. 1996. Sewall Wright meets artificial life: the origin and maintenance of evolutionary novelty. *Trends in Ecology and Evolution* **11**: 478–482.

Turchin, P. 1998. *Quantitative analysis of movement: measuring and modeling population redistribution in animals and plants.* Sinauer Associates, Sunderland.

Turing, A.M. 1952. The chemical basis of morphogenesis. *Philosophical Transactions of the Royal Society* **B 237**: 37–72.

Uchmański, J. and V. Grimm. 1996. Individual-based modelling in ecology: what makes the difference? *Trends in Ecology and Evolution* **11**: 437–441.

van Damme, J.M.M. 1985. Gynodioecy in *Plantago lanceolata* L.V. Frequencies and spatial distribution of nuclear and cytoplasmic genes. *Heredity* **56**: 355–364.

van Herwaarden, O.A. 1997 Stochastic epidemics: the probability of extinction of an infectious disease at the end of a major outbreak. *Journal of Mathematical Biology* **35**: 793–813.

Vance, R.R. 1985. The stable coexistence of two competitors for one resource. *American Naturalist* **126**: 72–86.

Vasudevan, K., W.G. Wilson, and W.G. Laidlaw. 1991. Simulated annealing statics computation using an order-based energy function. *Geophysics* **56**: 1831–1839.

Vincent, T.L. and H.R. Pulliam. 1980. Evolution of life history strategies for an asexual annual plant model. *Theoretical Population Biology* **17**: 215–231.

Wilson, W.G. 1998. Resolving discrepancies between deterministic population models and individual-based simulations. *American Naturalist* **151**: 116–134.

Wilson, W.G., S.P. Harrison, A. Hastings, and K. McCann. 1999. Exploring stable pattern formation in models of tussock moth populations. *Journal of Animal Ecology* **68**: 94–107.

Wilson, W.G. and W.G. Laidlaw. 1992. Microscopic-based fluid flow invasion simulations. *Journal of Statistical Physics* **66**: 1165–1176.

Wilson, W.G., W.G. Laidlaw, and K. Vasudevan. 1994. Residual statics estimation using the genetic algorithm. *Geophysics* **59**: 766–774.

Wilson, W.G. and R.M. Nisbet. 1997. Cooperation and competition along smooth environmental gradients. *Ecology* **78**: 2004–2017.

Wilson, W.G. and S.A. Richards. In Press. Evolutionarily stable strategies for consuming a structured resource. *American Naturalist* **155**.

Wilson, W.G., A.M. de Roos, and E. McCauley. 1993. Spatial instabilities within the diffusive Lotka–Volterra system: individual-based simulation results. *Theoretical Population Biology* **43**: 91–127.

Yamamura, N. and N. Tsuji. 1987. Optimal patch time under exploitative competition. *American Naturalist* **129**: 553–567.

Zeigler, B.P. 1977. Persistence and patchiness of predator–prey systems induced by discrete event population exchange mechanisms. *Journal of Theoretical Biology* **67**: 687–713.

Index